APPLIED CLIMATOLOGY

AN INTRODUCTION

APPLIED

OXFORD UNIVERSITY PRESS

CLIMATOLOGY

AN INTRODUCTION

SECOND EDITION

JOHN F. GRIFFITHS

Texas A&M University

Oxford University Press, Walton Street, Oxford OX2 6DP

OXFORD LONDON GLASGOW NEW YORK
TORONTO MELBOURNE WELLINGTON CAPE TOWN
IBADAN NAIROBI DAR ES SALAAM LUSAKA
KUALA LUMPUR SINGAPORE JAKARTA HONG KONG TOKYO
DELHI BOMBAY CALCUTTA MADRAS KARACHI

ISBN 0 19 859925 0

First edition 1966
Second edition 1976
Reprinted 1978

The photograph on the title page is reproduced by kind
permission of the Royal Meteorological Society

Text set in 10/11 pt IBM Press Roman, printed by photolithography,
and bound in Great Britain at The Pitman Press, Bath

TO JOAN

PREFACE TO THE FIRST EDITION

All of us are very much aware of the fact that climate and the weather play an important role in our every-day activities. Before leaving for work in the morning we usually listen to the weather forecast for the day to learn if an umbrella or overcoat is called for; can we prepare for that outdoor picnic; where should we go for a sunny summer holiday? Over the past twenty years the interest of the meteorologist has become increasingly focused on this type of problem – of a very practical and applied nature.

Applications of climatology are not discussed in most meteorological textbooks and are given only cursory mention in some of the books on general climatology. A few texts exist which cover one specific branch of the applied field, for instance on agricultural meteorology, but no publication deals with such a large number of applications. This book has been written to introduce interested readers to some of the many challenging problems of living with the weather.

The first part of the text deals with pure climatology and explains the famous Köppen classification, and gives selected temperature and rainfall figures for many stations around the world. The second part considers some of the ways in which climate affects other fields, such as soil, the human body, architecture, and many others. This book is intended to act only as an introduction to the complex and rapidly developing science of applied climatology. It is hoped it will stimulate more people to learn to live sensibly with the weather and, perhaps, will encourage a few to devote their talents to tackling the many problems that are still unsolved.

1966   JOHN GRIFFITHS

PREFACE TO THE SECOND EDITION

This revised edition introduces a simple climatic classification without stressing the Köppen system. In addition there are new chapters dealing with two facets of growing interest, urban climates and climatic change, both aspects with important practical implications. It should be emphasized that the book is meant as an introductory text and one around which, with the help of the references, the lecturer can weave his own path of specialization or generalization.

1976   JOHN GRIFFITHS

CONTENTS

1. THE GROWING SCIENCE OF APPLIED CLIMATOLOGY   1

2. INSTRUMENTS AND MEASUREMENTS IN
   APPLIED CLIMATOLOGY   3
   2.1 The role of observations   3
   2.2 The accuracy of observations   3
       2.2.1 Radiation ● 2.2.2 Sunshine ● 2.2.3 Light ● 2.2.4 Cloud ● 2.2.5 Temperature ●
       2.2.6 Precipitation ● 2.2.7 Humidity ● 2.2.8 Air movement ● 2.2.9 Pressure ●
       2.2.10 Evaporation
   2.3 Presentation and processing of data for practical use   8

3. ELEMENTS OF APPLIED CLIMATOLOGY   10
   3.1 Introduction   10
   3.2 Radiation   10
       3.2.1 Introduction ● 3.2.2 Astronomy ● 3.2.3 Depletion and scattering ●
       3.2.4 Long-wave and net radiation
   3.3 Sunshine, light and cloudiness   12
       3.3.1 Sunshine ● 3.3.2 Light ● 3.3.3 Cloudiness
   3.4 Temperature   13
       3.4.1 Introduction ● 3.4.2 Annual cycle ● 3.4.3 Diurnal cycle ● 3.4.4 Altitude effects ●
       3.4.5 Soil temperatures
   3.5 Precipitation   15
       3.5.1 Introduction ● 3.5.2 Annual values ● 3.5.3 Monthly values ● 3.5.4 Diurnal variation ●
       3.5.5 Altitude variation ● 3.5.6 Drought ● 3.5.7 Snow
   3.6 Humidity   17
       3.6.1 Introduction ● 3.6.2 Parameters ● 3.6.3 The dew point
   3.7 Air movement   17
       3.7.1 Introduction ● 3.7.2 Secondary flow ● 3.7.3 Variation with altitude
   3.8 Evaporation and evapotranspiration   19
       3.8.1 Evaporation ● 3.8.2 Evapotranspiration

4. OCEAN CURRENTS AND AIR MASSES   21
   4.1 Introduction   21
   4.2 Ocean currents   21

4.3 General circulation   21
4.4 Air masses   23
4.5 Fronts   25
4.6 Mesoscale disturbances   25

5. THE CLIMATE OF THE STANDARD CONTINENT   26

5.1 Introduction   26
5.2 The climatic factors affecting the standard continent   26
5.3 Temperature patterns on the standard continent   27
5.4 Precipitation patterns on the standard continent   27
5.5 Effect of shape and topography   29

6. CLIMATIC CLASSIFICATIONS   30

6.1 Introduction   30
6.2 Classification details   30
6.3 The tropics (A zone)   30
    6.3.1 Hot, wet (A1) ● 6.3.2 Hot, short dry spell (A2) ● 6.3.3 Hot, long dry spell
6.4 The subtropics   32
    6.4.1 Warm, uniform rain (BU) ● 6.4.2 Warm, summer rain (BS) ●
    6.4.3 Warm, winter rain (BW) ● 6.4.4 Warm, desert (BF)
6.5 The short-winter zone (C)   33
    6.5.1 Short winter, uniform rain (CU) ● 6.5.2 Short winter, spring rain (CV) ●
    6.5.3 Short-winter, summer rain (CS) ● 6.5.4 Short winter, winter rain (CW) ●
    6.5.5 Short winter, desert (CF)
6.6 The long-winter zone (D)   34
    6.6.1 Long winter, uniform rain (DU) ● 6.6.2 Long winter, summer rain (DS) ●
    6.6.3 Long winter, winter rain (DW)
6.7 Cold zone (E)   34
6.8 Special climates   34
    6.8.1 Mountain climates ● 6.8.2 Marine climates ● 6.8.3 Continental climates
6.9 Common classifications   35

7. MICROCLIMATES   49

7.1 Introduction   49
7.2 Radiation, sunshine, light and temperature   49
7.3 Humidity and evaporation   50
7.4 Air flow and precipitation   50
7.5 Modifications of climate or weather   51
    7.5.1 Windbreaks ● 7.5.2 Artificial stimulation of rain ●
    7.5.3 Reduction of evaporation ● 7.5.4 Frost prevention

8. CLIMATE AND SOILS   54

8.1 Introduction   54
8.2 Zonal soils   55
8.3 Intrazonal and azonal soils   55
8.4 Climate, soil and the standard continent   56

9. CLIMATE AND VEGETATION   57

9.1 Introduction   57
9.2 Forests and woodlands   57
9.3 Shrubland and grassland   58
9.4 Deserts   59
9.5 Climate, vegetation and the standard continent   59

10. CLIMATE AND AGRICULTURE   60

10.1 Introduction   60
10.2 Temperature   61
10.3 Moisture   62
10.4 Sunlight   63
10.5 Wind   63
10.6 Evaporation   64
10.7 Phenology   64
10.8 Climate and crop relationships   65
10.9 Artificial environments   65

11. CLIMATE AND FORESTRY   67

11.1 Introduction   67
11.2 Radiation   68
11.3 Temperature and humidity   68
11.4 Wind and precipitation   69
11.5 Stand patterns of climate   70
11.6 Timber production and forest fires   70
11.7 Windbreaks   71

12. CLIMATE AND HUMANS   72

12.1 Introduction   72
12.2 Heat balance   72
12.3 Water balance   75
12.4 Balance of the body   75
12.5 Empirical estimates of physical feeling   76
    12.5.1 The effective temperature ● 12.5.2 The temperature–humidity index ●
    12.5.3 The strain index
12.6 Clothing insulation and clothing zones of the world   77
12.7 Climate and the home   79
    12.7.1 Food ● 12.7.2 Clothing ● 12.7.3 Cleaning ● 12.7.4 Gardening
12.8 Human health   81
    12.8.1 Direct effects (detrimental) ● 12.8.2 Direct effects (beneficial) ●
    12.8.3 Indirect effects ● 12.8.4 Miscellaneous effects

13. CLIMATE AND ANIMALS   84

13.1 Introduction
13.2 Mammals   84
    13.2.1 Mammals in a natural environment ● 13.2.2 Domestic animals   85

CONTENTS

13.3 Flying animals 88
    13.3.1 Birds ● 13.3.2 Insects
13.4 Aquatic animals 89
13.5 Reptiles 90
13.6 Soil dwellers 90

14. CLIMATE AND BUILDING 92

14.1 Introduction 92
14.2 Climate and the architect 93
    14.2.1 Thermal considerations ● 14.2.2 Ventilation and wind pressure ●
    14.2.3 Daylighting factors ● 14.2.4 Precipitation or dampness aspects
14.3 Climate and the site 95
14.4 Conditioning by climate and design 95
14.5 Indoor climate 96
14.6 Heating and cooling 96
14.7 An elementary climatic classification for housing 98
    14.7.1 Tropics ● 14.7.2 The warm or subtropical regions ● 14.7.3 Cool regions ●
    14.7.4 Cold regions

15. CLIMATE AND HYDROLOGY 101

15.1 Introduction 101
15.2 Water gain 102
15.3 Catchments and depth-area-duration relationships 102
15.4 Surface runoff and underground water 103
15.5 Engineering hydrology 104
    15.5.1 Storage reservoirs ● 15.5.2 Flood control reservoirs ●
    15.5.3 Forecasting river conditions ● 15.5.4 Open-channel hydraulics ●
    15.5.5 River floods

16. THE URBAN ENVIRONMENT 106

16.1 Introduction 106
16.2 Physical characteristics of the urban area 106
    16.2.1 Hydrologic changes ● 16.2.2 Thermal changes ● 16.2.3 Aerodynamic changes
16.3 The climate of the city 107
    16.3.1 The urban heat island ● 16.3.2 The dust dome ● 16.3.3 Precipitation ●
    16.3.4 Small heat variations
16.4 Pollution in the city 109
16.5 Hydrology in the city 110

17. THE CHANGING CLIMATE 111

17.1 Introduction 111
17.2 Climatic factors 112
17.3 Methods of assessing past climates 112
    17.3.1 The uniformitarian principle ● 17.3.2 Sedimentary or geomorphological evidence ●
    17.3.3 Calculations
17.4 The historical past 113
17.5 The present 113

17.6 Deductions concerning past climates   113
17.7 Effect of changes in causal parameters   114
        17.7.1 Solar radiation ● 17.7.2 Atmospheric composition
17.8 The greenhouse effect   115
17.9 The vertical lapse rate of temperature   115
17.10 Changes that may be occurring now   115
        17.10.1 The observational approach ● 17.10.2 The theoretical approach
17.11 How would climatic change affect us?   118
17.12 The future   118

18. CLIMATE AND INDUSTRY, COMMUNICATIONS AND TRANSPORT   120

18.1 Introduction   120
18.2 Industry and commerce   120
18.3 Power and communications   121
18.4 Transport   122
        18.4.1 Air ● 18.4.2 Water ● 18.4.3 Railways ● 18.4.4 Roads
18.5 Military operations   125
18.6 Miscellaneous   127

REFERENCES   128

INDEX   133

REFERENCES

Journal articles and books referred to in the text and cited in the
References are indicated by superior numerals, e.g.[8].

Chapter 1

THE GROWING SCIENCE OF APPLIED CLIMATOLOGY

All around us, every day, and in almost all that we do, climate plays a part. Often the part is an important one, sometimes it is trivial, but many of our activities and behaviours are influenced by the elements of the climatic aspect of our environment.

Millions of years ago when the seas covered the world it is thought that the first forms of life existed in the equable environment of the water. There were no great temperature fluctuations or moisture stresses to bear and the organisms could adapt themselves gradually to the slow changes taking place. When the time of hydrological and major mountain formation occurred the exposed rock was subjected to the thermal changes that, through a long process of contraction and expansion, caused pronounced weathering and denudation

Through fossil, geological and archaeological information it is clear that animals and early man had the same three courses open to them as exist for the animals and for us today — to adapt themselves to the environment, to modify enough of the environment to make a suitable micro-habitat, or to perish!

Astronomy may be the oldest science but surely meteorology must come a close second, for as soon as man began to travel he needed to know something of the elements. To the early sailors wind direction was all important and it is on record that 2 000 years ago an example of applied climatology led to a substantial financial gain — as it so often does today. The Greek sailing ships plied between southern Arabia and India with rich cargoes, but for the journey to the Mediterranean they had to offload and employ camel caravans to cross the territory of the Sabean Empire, the rulers of which, naturally, took great care to exact duty on this merchandise. Hippalus, in the first century A.D., noted that during the northern summer the winds persistently blew from the west, towards India, while in the winter they blew from the east, from India. Armed with this information the sailors scheduled their sailings accordingly and plied between India and Ethiopia, even up the Red Sea, so avoiding the Sabean customs completely.

Climatology also, as far as we are aware, developed with the Greeks. Some twenty centuries ago they divided the world into three zones, namely, torrid, intermediate and frigid. In the first, the winterless zone, the sun was overhead at some period during the year while, in the latter, the summerless zone, there was a period when the sun never rose. Here the subject of climatology, more correctly when in its descriptive phase it should be termed climatography, rested for many years. In the eighteenth and nineteenth centuries many naturalists and other observers began to collect and analyse data, generally for the specific small areas in which they were working.

During the period from about 1900 to 1940 climatology remained a mostly descriptive subject, mainly detailing the relationship of vegetation to climate and presenting its information as average values of the parameters of certain climatic elements, generally, temperature and precipitation. However, during that time, a few 'pioneers' began to appreciate the service that climatology could give to other disciplines. In the mid-1920s Geiger, in Germany, began to investigate systematically the climatic variations that occur within very small horizontal and vertical distances and thus the science of micro-climatology came into existence. Dorno, in Switzerland, began to study the effect of the climate on people, an application of climatological knowledge. Landsberg, in the USA, was instructing in the subject during the 1930s, but it was not until the Second World War that more than a few people came to appreciate the important role of climatology.

The war led to fighting and military operations in many climatological zones of the world, from the frozen seas of the Arctic, the severe winters in the Ukraine, the tropical deserts of North Africa to the humid jungles of Asia and the Pacific. It was found, often by bitter experience, that machines, materials and even men that could withstand some environments could not withstand others. The military personnel began to appreciate the need to give rigorous tests to all equipment sent to other climatic zones and they began also to examine human reactions and needs under climatic stress conditions.

When the war ended applied climatology was just in its infancy and, as so often happens, when the incentive of the demands of war waned so did the scientific drive. However, now that more people appreciated the role of climatology in their own particular field of interest they began conducting experiments in microclimatology and its applications. It was not until 1956 that a move was made to unite workers in the science when Tromp, in Holland, formed the International Society of Bioclimatology and Biometeorology. The Society really grew from interests in the field of human physiology and climate, it being in this limited sense that the term bioclimatology is often interpreted in the United States. Bioclimatology, however, is accurately defined as the effect of climate on all aspects of living, its specialized connotation with respect to human physiology or phenology, as it was used in a pioneering study by Hopkins in 1938, being incorrect.

In this mundane world things have to be proved

to 'pay off' if they are to be accepted with alacrity. What a difficult task this is to do in meteorology — how much is a good forecast worth? Who can estimate, and then put a value on, the lives and property saved by the excellent forecasting for hurricanes Hattie (destroying Belize, British Honduras) and Carla (hitting the Texas coast)? How many lives are saved each day by accurate flight information, how many acres of crops are saved if a frost warning is issued in sufficient time, how much comfort is dispensed if the electricity companies balance their loads when warned of approaching cold weather? However, the subject was thought to be of such importance that the National Industrial Conference Board held a special 'Business of Weather' session in Chicago in November, 1962, and the four invited papers[1] give an interesting insight into the economics of applied climatology.

This book is not highly technical; but the undergraduate and graduate student or researcher can find here some of the present-day thoughts on applied climatology and it is hoped these will stimulate him to other investigations in this very new science. The book is in two main sections, the first, Chapters 2–8, dealing with climatology and microclimatology, while the last chapters try to cover the numerous disciplines to which climatology is applied.

Climatology is, of course, the central or integrating science for all disciplines in the field of applied climatology and the climatologist must now be instructed in the various methods by which he can be of the greatest general service to the public, 'man-in-the-street' and research worker alike, if applied climatology or bioclimatology is to advance at the rate needed in this time of population explosions, increasing land use and the search for better living.

Since the first edition of this book (1966) other, and in some cases, more technical, books have appeared that deal with some of the applications of climate to life.[2-7] It is heartening for the science of climatology that so much interest is now given to this practical facet.

Chapter 2

INSTRUMENTS AND MEASUREMENTS IN APPLIED CLIMATOLOGY

2.1 The Role of Observations

The whole subject of climatology, both pure and applied, is based upon numbers and, unfortunately, people are apt to use these numbers without questioning their meaning or significance.

There exists a tendency, when studying an aspect of applied climatology, to accept that the values recorded at the nearest climatological site are closely related to the conditions at the place under investigation where a different microclimate may exist. Before one can judge whether or not this is so one must know the two sites and, equally important, must know the instruments that have been used, their limitations, their correct and actual exposure and, if recorders are used, whether these are suitable for the element to which they are connected.

Another problem is that concerning accuracy — does the instrument give too little or too much? Both these possibilities can cause great confusion when looking for relationships. Each climatic parameter thought to be relevant should be considered in turn and the desirable accuracy predetermined wherever possible.

2.2 The Accuracy of Observations

Within the scope of this book it is impossible to make sweeping generalizations concerning all the climatic elements, but we can discuss shortly the major elements in turn. The reader interested in fuller details of instrumentation in environmental studies is referred to recent publications by Platt and Griffiths[1] and Monteith.[2] Standard

Meteorological instruments are fully discussed by Middleton[3] and HMSO.[4]

2.2.1 *Radiation*

Radiation is not a standard climatological element and it is measured on a continuous basis at only a few, about 600, stations around the globe. The coverage of stations is generally quite inadequate for the construction of maps, except in the USA, a few countries of Europe and the Republic of South Africa.

The accuracy of the measurements made is usually of the order of 5 per cent over short periods, about a few hours, and improves to about 3 per cent for periods of a day, provided good equipment and servicing are maintained. The instruments are generally rather expensive and need an experienced technician to ensure their continuous functioning. Few models of sufficient accuracy are self-contained, that is, most need some form of external recorder. When these are used there arises the lengthy process of summing the records over the periods required. This can be overcome by employing a special type of integrating mechanism but its use tends to increase the cost until, finally, apparatus that records and integrates the short wave, or solar, radiation costs in the region of £1 400. Few services can afford to set up networks of these stations.

Instruments can be obtained that are much cheaper but, it is fairly true to say, the cheaper the instrument the less its accuracy, although a cheap model may prove satisfactory for the measurement of radiation over long periods, say

1 month or more. It is this variation with instrument that makes it essential to know just what type of apparatus was used for any particular set of measurements.

On the market there are standard models for the measurement of global (sun and sky) short-wave radiation, direct solar radiation and sky, or

Fig. 2.1 Radiometer and shadow ring

diffuse, radiation. The latter model usually employs a shadow ring (Fig. 2.1) to shield out the direct radiation so that its use necessitates correction factors that are not known with great accuracy. The direct measuring instrument is normally mounted on a heliostat and follows the sun's movements, the acceptance angle of the element being quite small, about 2°. However, even such a small angle does not accept only the direct solar rays and corrections are applied for accurate work.

There are available on the market filters so that certain sections only of the spectrum can be measured. Such filters do not have exact cut-off limits and so are only approximations to the absolute values. They can prove extremely useful when measuring the response of animals or plants to certain types of energy rays.

Most models available need some form of external power, but a few are self-contained and record on clockwork-driven charts. A number of models take only spot readings and cannot be made to record.

So far we have mentioned only short-wave radiation, defined in meteorology as that band between 0·3 and 5 μm. However, the measurement of terrestrial radiation, mainly in the infra-red band above 5 μm, is also very important. Much difficulty has been experienced in obtaining some

substance that would shield the element and yet allow the passage of the radiation. The problem is by no means completely solved and, although there are many different models on the market, none is ideal. As the interest in the measurement of long-wave radiation has developed only during the last decade there are but a handful of stations that make continuous records of this element and, in order to obtain representative values, the investigator normally has to make measurements for himself.

The interrelationship between short- and long-wave radiation leads to the study of the heat balance, or flux relationships. This subject is, of course, even newer than the long-wave study, and very few measurements have been made. Any one interested in such measurements will do well to refer to the current literature for the latest and ever-occurring developments of this problem.

2.2.2 *Sunshine*

The standard sunshine instrument, the Campbell-Stokes model, was developed during the last century and consists of a glass sphere focusing the solar rays on to a sensitized paper, thereby causing a burn to appear whenever the heat rays are powerful enough. In effect the instrument is almost a radiation instrument but it simply records on an 'on-off' or 'yes-no' routine. Common practice is to take the times during which burning occurs as periods of 'bright sunshine'; note that the term 'bright' is not defined.

In general, the sun is not sufficiently powerful to cause a burn to appear before it has risen about 5° above the horizon so that the hours of sunshine, such as appear in the newspapers or in climatological records, are often underestimates.

Other sunshine recorders exist, the main one being that used in the USA where the heat of the sun causes mercury in a thermometer to expand so that a switch is closed. Again, it has the disadvantage of not recording until a certain heat threshold is reached, a threshold approximately the same as that of the Campbell-Stokes model.

Due to the threshold value being a heat and not a light measurement, it is possible to have sunlight and no burn, which occurs when a low sun is covered by a thin high cloud, or no sunlight and a burn, when the noonday sun is covered yet there is sufficient concentrated diffuse radiation to burn the sensitized paper. These factors can also lead to errors in the recorded measurements but, over long periods, they tend to cancel out.

A very real source of error is the interpretation of the burnt charts. It was found that different individuals obtained different values of 'sunshine hours' from the same chart and this has caused the World Meteorological Organization to stipulate definite rules for the analysis of these charts. As sensitized paper is used it is also essential to ensure that the sensitization and thickness of paper are uniform, because stocks from different manufacturers can vary.

Finally, it is essential to mount the apparatus accurately because any incorrect mounting, such as the axis not pointing to the celestial north pole, non-alignment with the meridian or shading occurring from any object such as a flag-pole or a mountain, will cause systematic errors that may be impossible to correct.

2.2.3 Light

As has been pointed out above, there is some confusion between the term 'sunlight' and radiation. A similar confusion often appears between 'light' and 'heat'. The term 'light' is reserved exclusively for the band between about 0·4 and 0·7 μm, that is, the band to which the normal human organ of vision is sensitive.

Light is not measured as a standard climatological parameter and readings, on a continuous basis, are very sparse. The measurement is normally made with some kind of photoelectric cell using a filter that reduces its response to that of the human eye. The filters, when left exposed for long periods, collect deposits and have to be cleaned often. The photoelectric cell must be calibrated often as ageing occurs.

2.2.4 Cloud

Cloud is normally not recorded by instruments but is estimated by an observer on the spot. The observations are made a few times each day, varying from every hour at main stations to once a day at smaller observatories. The readings, made in eighths of the celestial hemisphere covered, are averaged and the mean is taken to be the average cloud cover. In recent years the fish-

eye lens and dome shaped reflectors have allowed photographic records to be made.

The disadvantages of such a system are obvious, especially when it is realized that most stations send in only two or four cloud estimates per day. In some places, at certain seasons, clouds show a marked diurnal variation and this may be missed or accentuated by the selection of times of observation. Night-time observation is difficult and most stations depend upon day-time estimates.

Pictures from satellites will, of course, assist in giving an integrated picture over a large area as opposed to the present method wherein an observer's vision covers a small area. In addition the satellite is able to use infra-red sensors that can locate cloud tops by radiation temperatures.

2.2.5 Temperature

This element is normally measured with accurate thermometers or recorders and, provided the instruments are well looked after, will yield sufficiently accurate results, sufficient that is for most applied investigations. The exposure of these instruments has to be carefully arranged.

The sensitive element of the instrument is of a finite size and so will tend to absorb heat if exposed to radiation, giving it too high a temperature during the day and too low a temperature at night due to its interchange of radiation with the cool sky or surroundings. Also, there must be sufficient flow of air across the bulb or element so that stagnation of hot or cold air does not occur. For these reasons it is standard practice to expose a thermometer or thermograph in an instrument shelter or screen with louvred sides. This ensures that no direct radiation falls upon the instrument while there is also sufficient air flow across the apparatus. Because of the large changes of temperature that take place within the lowest few feet of the atmosphere, especially in and around natural and artificial objects, it is usual to expose the instruments at a height of about 1·5 m to obtain a typical air temperature.

It is essential to know if such precautions have been taken when using data concerning air temperatures. It is normally assumed that air temperature measurements reported by all meteorological services have been made in instrument shelters.

When dealing with temperatures at other than screen height, as for instance in a crop field, or at a height of, say, 30 cm, the same type of precautions must be followed, that is, the instrument must be shielded from radiation but exposed to air flow. This can become very difficult to achieve in practice, if one is not to disturb the natural conditions. Recourse is therefore often made to the use of small thermo-electric devices, such as thermocouples, resistance thermometers or thermistors. These instruments have many possible sources of error and great care must be taken in their use. When accepting measurements made by other workers it is essential to know how the apparatus was set up, its shielding, calibration methods and the recorders or galvanometers used. The infra-red thermometer is discussed in 3.2.4.

Soil temperatures are easier to measure than air temperatures, because there are no problems of radiation or air flow with which to contend. It is necessary to ensure that the element is in close physical contact with the soil at the depth required, for air pockets within the soil can give unrepresentative temperatures. Care must be taken to see that there is little conduction of heat to the bulb of the thermometer through the stem or the case. When thermo-electric means

best to attach the elements to a post driven into the bed of the pond.

2.2.6 *Precipitation*

Rainfall measurements were apparently first made in India during the fourth century B.C. and, in those regions where water is in short supply at some time of the year, close attention has been paid to the amounts of rainfall for many centuries.

Rainfall is generally measured by collecting the water in some form of container and converting this amount to an equivalent depth. The size of the container and the exposure vary appreciably from country to country but, generally, the water is collected in a circular container, from 12·5 cm to 30 cm diameter, exposed at a height of from 45 cm to 100 cm above the soil surface. The container must of course be remote from any shielding effects, such as a tree or a building.

Most rain gauges are of the type that need to be read by an observer. Readings are usually taken once a day so that no idea of the diurnal pattern of the rainfall can be obtained. Some gauges are

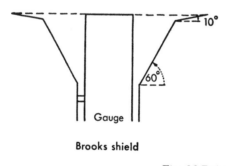

Brooks shield

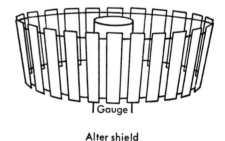

Alter shield

Fig. 22 Rain gauge shields

are used it is necessary to check from time to time to note if any galvanic effects are present due to the chemicals in the soil affecting the wires.

Water temperatures are often taken only roughly, and it is advisable to use small thermo-electric elements to obtain accurate readings. The difficulty here is to ensure that the element is at the required depth, an especially difficult problem for bodies of water with pronounced wave motion. On ponds, pools and the like it is

of the recording variety; these chart the amount within the container and enable a record of diurnal distribution and intensity to be calculated.

A rain gauge naturally presents an obstruction to the air flow so that a rising current on the windward side and a downward current on the leeward side will occur. To ensure a horizontal flow across the gauge, cone-shaped shields are often used (Fig. 2.2), but these can cause blocking of the gauge if snow fall is heavy. Because the air speed normally increases with height (see 3.7.3) there is a loss of 'catch' of rain the greater the height of the rim above the ground.

It is standard procedure to accept as an estimate of the rainfall in an area the amount collected by one gauge. If the rain gauge is of the large type (20 cm diameter) it collects from 1 part in 3×10^7 of a square kilometre a very, very small sample. In many parts of the world, even on land, it is common to have only one gauge every 1 000—3 000 square kilometres, so that to consider the catch as representative is not likely to be very accurate. As the time interval is increased, the accuracy of the sample, in a region of fairly uniform terrain, also increases, and the mean annual total from such a gauge may give a good estimate of the areal annual fall, whereas it would not give a good idea of the average fall from a single storm of a few minutes' or hours' duration.

Rain gauges must be kept in good condition and it is useless to have a leaking gauge, although such are not unknown in the remoter parts of the globe.

The exposure of the gauge is fundamental. Often a gauge is put in a particular position from considerations of convenience, either of the space available or to suit the observer, and no regard is paid to the proximity or obstacles that may influence the catch from some or all directions. The readings from a well-positioned gauge are sometimes influenced by the growth of a tree, previously too small to cause any interference with the catch in the instrument. The use of radar in estimating rain amounts is still in the experimental stage, but the accuracy of the digitized technique is being improved frequently.

2.2.7 *Humidity*

Normally when the term humidity is used it is the relative humidity of the air that is inferred. The relative humidity is the ratio, in per cent, of the amount of water vapour in the air to that which the air would hold if saturated at the same temperature. The amount of moisture that the air can hold is a function of the air temperature so that relative humidity at a station is also related to the air temperature. This relationship is often so pronounced that the relative humidity pattern is almost the exact inverse of the air temperature pattern. Because of this it is essential to know the air temperature at the same time as the relative humidity, a fact it is not always possible to discover from published data.

A better parameter to use is the absolute humidity, or the measure of how much water vapour is held within a certain volume of air. This can be obtained simply from the dew point temperature, that is, the temperature at which water will begin to deposit out if the parcel of air is cooled gradually. This can be measured directly or calculated from the air temperature and the relative humidity or from the air temperature and the wet-bulb temperature (the temperature to which the air can be cooled by the evaporation of water into it).

Most humidity measurements are made from either a hair hygrograph or a psychrometer.

The functioning of the hair hygrograph depends upon the fact that human hair alters in length as relative humidity changes. The movement of the hair is used to activate a pen that marks on a rotating chart. The instrument is usually large and must be shielded from the radiation when in use. The psychrometer, using two thermometers, one an ordinary model the other being covered with a moistened sleeve, must have a sufficient air motion across the bulbs to ensure correct readings. This means that, in the relatively calm conditions of the lower layers of the atmosphere, there must be artificial ventilation. Such an induced air motion causes mixing in the lower levels and the result is then unrepresentative. The modern method is to use some of the thermo-electric means mentioned under *Temperature* (2.2.5), but these will also require ventilation, although of less intensity.

2.2.8 *Air movement*

Air movement must be considered under two sections: direction and speed.

Wind direction, which often varies greatly, can either be estimated by smoke drift, measured by a vane or recorded by mechanical or electrical means. Provided the exposure is suitable for the investigation undertaken, all methods are satisfactory.

Wind speed may prove more difficult. The normal meteorological exposure of an anemometer is at a height of at least 10 m because below this the influence of the immediate surface is dominant. Within the surface layers the change of wind direction can be such that some types of anemometers, such as the windmill model, can reverse their direction of revolution and give completely erroneous readings. One of the major difficulties is the low wind speeds encountered in these surface layers, necessitating the use of sensitive anemometers. These are, of course, more expensive and need careful handling.

However, in general, the making of air flow measurements is not so difficult as for many of the other climatic elements and there is little need to dwell further upon this aspect.

2.2.9 *Pressure*

Pressure is not, strictly, a climatological factor, its variations only being experienced through other elements such as air movement. However, the need sometimes arises to make pressure measurements.

The two types of pressure-measuring instruments most frequently used are the barometer, using a mercury column, and the aneroid cell, using an evacuated chamber. The former is much more accurate but cannot be moved from site to site and is usually installed permanently in an observatory. The aneroid cell, which can be made to record, is sensitive but not so accurate. It is generally sufficiently accurate for applied climatological investigations and if the microbarograph, using a series of cells, is employed more sensitivity is obtained. The instrument must be handled carefully and exposed within a room away from radiation and wind effects.

2.2.10 *Evaporation*

Evaporation, in climatology, is meant to be the estimate of the evaporation from a freely-exposed open water surface. The measurements that are found in the literature are measurements made from many different types of pans which vary in size, depth, amount of water, material of construction and exposure. The variation is so great that it is virtually impossible to transform the data from one pan into equivalent amounts from another pan except on a long time scale, such as yearly or monthly. Using an evaporation pan it is not easy to assess what meteorological parameter is being measured and evaporation poses some of the most complex, yet pressing, problems in meteorology today.

2.3 Presentation and Processing of Data for Practical Use

Climatic data are taken, normally, at a fixed site so that three basic problems arise.

(1) Are the data at that site presented in a usable form for the problem on hand?
(2) Are the data at the site at a certain time representative of those data at a different time at the same site?
(3) Are the data at the site representative of those data at another site in that same climatic zone?

(1) The presentation of the data must depend on the specific problem being investigated, but it is generally insufficient to have the parameters presented only as mean values, the commonest form of presentation. The best approach is to use some method relating values of the parameter to the frequency of their occurrence; for example, what is the percentage occurrence of temperatures above 30 °C, or what is the probability of more than 5 cm of rain during January? The distribution can be either derived simply from the past measurements or, and this is much better, examined statistically in order to find the simple distribution of best fit. The latter method helps to eliminate the possible sampling errors and such presentations are now becoming more frequent but, because of the work involved, they still do not appear often in print.

Many climatic data exist – in fact, almost an embarrassing number. They can seldom be sufficiently processed, even with punched card techniques, and the investigator must condition himself to the realization that he may personally have to arrange for the climatic parameters that interest him to be extracted and analysed.

(2) This problem poses the whole question of trends, linear or periodic. Many climatic investigations have been published using either Fourier or spectrum analysis to show the existence of certain cycles in climatic elements or weather phenomena but, although many cycles of statistical importance have been promulgated, it is unlikely that those of practical importance and significance allied to a corresponding physical interpretation constitute more than a very small percentage. One must be very careful to use statistics wisely and not to accept

blindly the fact that a cycle exists because this is inferred by the arithmetic. It must be realized that to substantiate a cycle of n years it is necessary to have approximately $4n$ years of data.

(3) This is a problem that can be resolved only by 'on the sites' inspection and knowledge. It is usually a good thing to run a trial comparison between climatic parameters at the two sites, but if this proves impossible the microclimates should be estimated by a consideration of such aspects as radiation patterns, air flow, vegetation and soil.

Chapter 3

ELEMENTS OF APPLIED CLIMATOLOGY

3.1 Introduction

A distinction must be drawn between the factors and elements of applied climatology. A factor is a determining cause of a phenomenon, an aspect that gives rise to a certain effect, while an element is a component part of that effect.

The major elements of applied climatology are radiation, the sunshine, light and cloudiness group, temperature, precipitation, humidity, air movement and evaporation. A thorough study of the parameters of these elements leads to a complete climatographic picture of an area, but does not explain the climatic causes.

This chapter is concerned with each of the elements in turn, trying to discuss the physical reasoning behind their basic patterns so that the researcher may then be aware of the variations from site to site or from day to day.

3.2 Radiation

3.2.1 *Introduction*

The sun is the source of all our energy whether it be in the form of coal, oil or direct heat. Its variations during the seasons give rise to differences in temperature, pressure, ocean currents, air masses and other major synoptic phenomena. On a long-term basis the incoming and outgoing energy of the earth and its atmosphere balance, but over a short time this is not so, and for any particular day there is an imbalance that is taken care of by storage or loss of heat in the ground, water or atmosphere.

Because of its tie-in with so many elements radiation is the most important of the climatic elements and it is unfortunate that it is one of the least measured. This fact means that instead of being able to study the cause of variations we have

often to be content with a study of the effects, such as temperature changes.

3.2.2 *Astronomy*

The sun, at an average distance of 150×10^6 km from the earth, gives out radiation at an approximately constant rate, equivalent to about 2·0 cal/cm² min on a surface normal to the beam at this mean distance. Because the earth's orbit is elliptical and not circular, the distance variation causes a change of intensity of about ±3 per cent in this 'solar constant'.

The angle that the solar beam makes with the equatorial plane is called the solar declination, d, and varies from +23°27′ to −23°27′. Owing to the fact that the earth's path is elliptical the sun does not appear to move in a steady rate across the sky and the concept of a heavenly body, called the mean sun, revolving at a constant rate, has been introduced. The difference between the times calculated from this mean sun and from the actual sun is called the equation of time, E. A knowledge of d and E enables the altitude and azimuth of the sun at any latitude at any time of the year to be calculated.

3.2.3 *Depletion and scattering*

On arrival at the atmosphere, the insolation (*in*coming *sol*ar *radiation*) is depleted and scattered by dry air molecules, water vapour, water droplets and dust, among other things. The amount of this depletion depends, naturally, upon the degree to which these are present in the atmosphere and also upon the length of the beam's path through the atmosphere. This path length, or optical air mass, is related to the solar altitude and is approximately proportional to the secant of this angle. The amount of radiation received on a surface is also proportional to the sine of the angle made between the solar beam and the surface so that as this becomes small the radiation received is correspondingly less.

that body. It has been shown that, to a good approximation, both the sun and the earth behave as black bodies. Thus, knowing the earth's temperature over a certain area we can calculate the approximate amount of radiation it is emitting and also the spectral distribution of such radiation. Also,

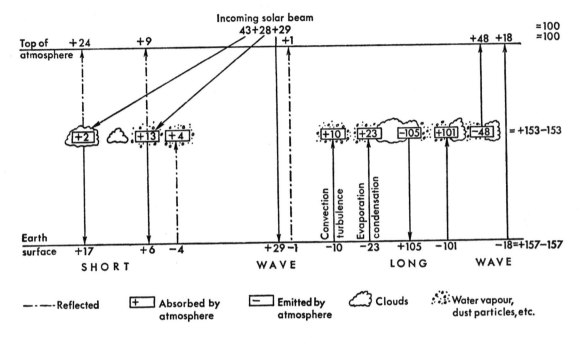

Fig. 3.1 The radiation balance of the earth

Fig. 3.1 gives some idea of the average values of the radiation balance and, though it does not represent any one place or any one day, it gives an idea of the values involved.

The intensity of radiation from the sun varies with the wavelength (Planck's law), the maximum intensity of the undepleted beam being in the yellow-green band. The wavelengths are not equally scattered, the general rule being that the shorter the wavelength the greater the scattering, providing the scattering particle is of small size compared to that wavelength. This fact means that the blue light is scattered much more than the red, hence the blueness of the sky.

Planck's law enables us to show that the amount of energy given off by the perfect 'black body', a hypothetical substance absorbing all the electromagnetic radiation falling on it, is proportional to the fourth power of the absolute temperature of

there is a simple relationship between the radiative temperature and the wavelength of maximum intensity of radiation. For example, the maximum intensity of the earth's radiation is at about 10 μm while that of the sun is at about 0·5 μm.

3.2.4 Long-wave and net radiation

Long-wave, or terrestrial, radiation is normally thought of as the radiation between wavelengths of 5 μm and 50 μm. As the earth radiates like a black body and its temperature is around 300 K, Planck's law enables us to calculate, and measurement confirms, that the major portion of terrestrial radiation is within this band, with a maximum at 10 μm.

The part played by the long-wave radiation in the long-term balance of the atmosphere is given in Fig. 3.1, and it can be noted that the values of the earth–atmosphere and atmosphere–earth radiation are both approximately equal to the incoming solar undepleted beam.

Kirchhoff's law states 'at a given temperature the ratio of the absorptive power to the emissive power for a given wavelength is the same for all bodies'; hence a good absorber is a good radiator and vice versa. For short-wave radiation, of which we can see a large percentage, it is possible for us to estimate roughly, whether a surface is a good or bad reflector, but in the long-wave radiation band this is not so because it is outside the limits of human vision. Fortunately most substances emit about 90–95 per cent of the black body radiation in the long-wave region but it is necessary to consult tables or carry out some observations in order to check the long-wave reflectance or emissivity of substances.

Absorption in the long-wave region, due to the presence of water vapour and carbon dioxide, is pronounced and, within the atmosphere, the only band relatively unaffected by absorption is the so-called 'window' between 8 and 13 μm. It is this window that causes the direct loss of 18 units to outer space in Fig. 3.1.

Because of the Planckian relationship that, for a perfect black body,

$$\text{emitted radiation} = \sigma T^4$$

where T = temperature (K),
σ = 82 x 10^{-12} cal/cm^2 min K^4,

it is possible, knowing or estimating the emissivity and the temperature, to calculate the outgoing terrestrial radiation. Conversely, by measuring the outgoing radiation it is possible to calculate the surface radiative temperature. This latter method has been used in remote areas to obtain temperatures from measurements made from low-flying aircraft and Tiros satellites. These infra-red thermometers actually measure the radiation in the 8–13 μm band and, by assuming a constant emissivity, can give a direct temperature reading.

3.3 Sunshine, Light and Cloudiness

These three elements of climate are all intimately related to the radiation.

3.3.1 Sunshine

Sunshine, as we have noted earlier, is very poorly defined and the investigator has to accept that the instrument used is making an actual measurement of this element.

Many relationships between sunshine and radiation have been propounded because more sunshine records than radiation records are available. Perhaps the best of these equations[1] is

$$Q/Q_0 = 0.29 \cos \varphi + 0.52\, n/N$$

where Q = total radiation on a horizontal surface at latitude φ
Q_0 = the same as Q in the absence of an atmosphere
n = the actual duration of sunshine
N = maximum possible hours of sunshine.

It must be realized that N should actually be reduced by a factor depending upon latitude and solar declination because otherwise the ratio n/N can never equal unity (see 2.2.2).

This equation is best used only over long periods, say 10-day means at least, since single days can give values at great variance with the equation.

3.3.2 Light

As the average solar spectrum distribution of energy is adequately known, as is the sensitivity of the average human vision, it is possible to relate the two variables and say, for instance, that the solar constant of 2·00 cal/cm^2 min is equivalent to about 145 kilolux. As the depletion that occurs alters the spectral distribution of the solar beam the relationship does alter but, for most practical purposes, it is possible to assume that the above holds true as a linear equation. It becomes in appreciable error for overcast conditions.

Light meters or photo-electric cells are often used in order to obtain some idea of the intensity but care must be taken to ensure that the angle of exposure is known and that the cell has a known calibration, because some show definite aging characteristics, especially if left exposed for long periods of time or if they are improperly sealed against moisture and dirt.

3.3.3 Cloudiness

Cloudiness is, naturally, vaguely related to the sunshine but, because of the varying hours of sunshine and the fact that some cloudiness figures include night-time observations, the relationship is not perfect.

However, the equation

$$100 = S + C$$

where both sunshine, S, and cloudiness, C, are expressed as per cent holds quite well for long-period averages, such as yearly values, and also for summertime means when the sun is at a high altitude.

In the temperate zones cloud frequency presents a U-shaped distribution, that is, small and large amounts of cloud occur more freqently than the middle group. This is because the synoptic situation is generally either conducive or not conducive to the formation of cloud.

It is possible to derive equations expressing the number of clear days as a linear function of the cloudiness, similarly with the cloudy days, but the constants have to be calculated for each station. Equations also exist to reduce the undepleted radiation values to observed values by applying factors depending upon the cloud types, but these are also of doubtful use except over long periods of time.

3.4 Temperature

3.4.1 Introduction

Temperature is mainly a manifestation of radiation and, in nature, this means that it is related to the interplay between solar and terrestrial radiation, plus the physical characteristics of the radiated and radiating surfaces. This relationship, based on known mathematical laws, enables a calculation of temperature to be made, providing all the radiation factors and the physical characteristics are known. However, it is much easier to measure the temperature directly.

Temperature can vary by about 145 °C over the earth's surface. The highest on record is 58 °C at San Luis, Mexico, while the lowest is −87.5 °C at Vostok, Antarctica. The highest mean annual temperature is about 35 °C at Vallol, Ethiopia, while the lowest mean annual temperature is −58 °C recorded at the Pole of Cold, Antarctica.

Some of the temperature terms in common use are defined below.

Mean daily temperature: the mean of the 24 hourly values, but it is generally taken to be the mean of the maximum and minimum.

Mean monthly maximum: the average, over the month, of all the daily maxima.

Mean monthly minimum: the average, over the month, of all the daily minima.

Mean monthly temperature: the average of the daily means for that month.

Diurnal range: the difference between the maximum and minimum for the day.

Annual range: unless otherwise specified it is the difference between the mean monthly temperatures for the hottest and coldest months.

Interdiurnal change: the difference between the mean daily temperatures of two consecutive days.

3.4.2 Annual cycle

The annual cycle of air temperature in regions unaffected, or relatively uniformly affected, by cloud and precipitation is closely related to the solar declination, but with a time lag of about 1 month, as the values for London demonstrate (Fig. 3.2). In continental locations the lag is often from 2 to 4 weeks, and may be as great as 6 to 8 weeks in marine areas.

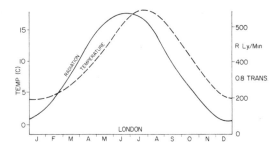

Fig. 3.2 Relationship between temperature and direct solar radiation on a horizontal plane, London. Radiation transmission: 0.8. 1 Ly = 1 cal/cm^2

However, when seasonal precipitation occurs, due both to the presence of increased cloud and the cooling effect of the rain, this simple relationship is often masked. For instance, Fig. 3.3 shows the typical pattern found in that part of India affected by the seasonal rains.

In marine climates the annual range is much less than in continental climates due to the modifying effect of the oceans, for instance, Jaluit (Marshall Is.) has a range of about 0.5 °C while Winnipeg (Canada) has 40 °C.

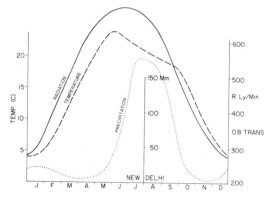

Fig. 3.3 Temperature, rainfall and solar radiation patterns for New Delhi. Transmission: 0·8.

3.4.3 *Diurnal cycle*

Like the annual cycle this pattern is also related to the solar radiation, if other factors do not interfere. The type of relationship is as shown in Fig. 3.4.

Again, a marine climate shows a smaller diurnal variation than a continental climate; Jaluit 6 °C, Winnipeg 13 °C.

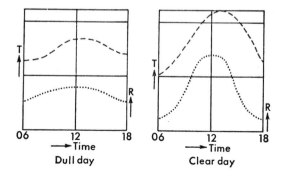

Fig. 3.4 Schematic relationship between daily variations of temperature, *T*, and radiation, *R*

The mean daily temperature is generally taken to be equal to the average of the maximum and the minimum for that day. This is not always correct but over long periods of time, such as a month, the approximation is usually good to within about 0·6 °C. The tendency with using this method is to obtain an overestimate of the mean.

3.4.4 *Altitude effects*

Within the free air the diurnal range decreases as the height above ground increases. Experiments on the Eiffel Tower[2] showed that the range was reduced to about 60 per cent of its 2-m value at a height of 200 m. Also, there is a marked delay in the time of the occurrence of the maximum and minimum, due to the effects of convection, turbulence and conduction. In the same experiment there was a delay of about 1·5 hours in the time of occurrence of the maximum and of nearly 1 hour in the minimum.

For stations at elevations above sea level, in which the temperatures are still taken within a shelter at about 1·5 m above the ground, there is a general lapse rate of about 0·55 °C/100 m. This compares with the theoretical adiabatic lapse rate for perfectly dry air of about 1 °C/100 m. An adiabatic process is a change of state of a system in which there is no transfer of heat nor mass across the system boundaries. The figure of 0·55 °C is only an average of many observations at stations all over the world and, for any given situation, it may range between 0·4 and 0·8 °C, or even more. There is also a tendency for the diurnal range to be greater at higher elevations because of the reduced blanketing effect of the thinner atmosphere.

In the first few centimetres above the earth's surface some very large gradients can exist. For example, the author has measured, using thermocouples, during the hot season in Southern Arabia.

(a) 71 °C at the surface, 38 °C at 1 m—a gradient of 2 800, dry lapse rate.
(b) 77 °C at surface, 49 °C at 5 cm—55 000, dry lapse rate.

These values are extremes but give some idea of the different orders of magnitude involved.

3.4.5 *Soil temperatures*

The pattern of temperature within the soil is dependent upon two main physical parameters, the heat conductivity and the heat capacity. These two regulate the temperature patterns that occur but they may vary as the soil is wetted or dried out. Since air is a poor conductor (0·003 cal/cm^2 min °C) but is transparent to radiation, whereas soils are better conductors but are opaque to radiation, it is to be expected that soil temperatures will exhibit different patterns from air temperatures.

Dry, sandy soils heat very rapidly at the surface during the day, due to the small heat capacity and poor conduction, but at a depth of a few centimetres the heating is much reduced (Fig.3.5). For a moist, loam-type soil the pattern is as shown in Fig. 3.5, the surface not heating as much while conduction is increased. At night the reverse occurs; the sand cools more rapidly than the loam due to the poor conduction of heat from below.

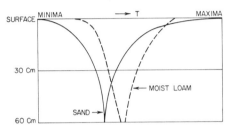

Fig. 3.5 Schematic diurnal pattern of temperature, T, variation with depth in sand and moist loam

To give some representative average soil temperature values, at about 3 m depth the annual range is about 3 °C, or less, at 6 m some 2 °C, at 10 m only 1 °C. In water, a substance transparent to radiation to a depth of many feet, at 20 m the annual range is still about 5 °C and is even 1 °C at 50 m.

As with the temperature pattern in the free air, there is a time lag caused by the heat flow through the soil. This is normally about 12 hours at 30 cm and about 6 months at 10 m. This means that at 30 cm depth the warmest time of day is around midnight and at 10 m the warmest time of the year is during the winter.

At a certain depth in the soil, depending upon its characteristics and local conditions, there is generally a mean annual temperature equal to that of the mean air (screen) temperature.

For a detailed account of the processes of night cooling and day-time heating the reader is referred to Brunt's textbook.[2]

3.5 Precipitation

3.5.1 Introduction

It has been pointed out earlier (2.2.6) how small a sample is the catch of a rain gauge and, because of this, the observed extreme values may well have been exceeded. At present, the greatest known annual rainfall average is 11·7 m on Mt Waialeale in Hawaii, although Cherrapunji (India) and Debundscha (Cameroons) have 10·8 m and 10·3 m respectively. The driest places in the world are Arica (Chile) and Wadi Halfa (Sudan), both of which have recorded only a few rains during tens of years of records. Cherrapunji has had over 26·4 m during a year and holds the record for the most in one month, 9·3 m. The greatest fall in 1 minute is 31 mm at Unionville, Maryland, and in 1 day, 187 cm at Cilaos, Reunion Island. There is an interesting empirical relationship between the world record rainfall R (cm), during time T (hours), given by

$$R = 40 \sqrt{T}$$

For precipitation to occur the presence of moist air is essential. If the air is too dry, precipitation can fall from a cloud and never reach the ground. The visible traces of precipitation that does not reach the ground are called virga. Once the assumption is made that a moist air mass exists, there are four main causes of precipitation. All of these causes have the effect of making the warm air ascend. On ascent the lower pressure leads to expansion and loss of heat, the resultant cooling means that less moisture can be retained and precipitation ensues.

(1) Orographic cause: Due to high land the air mass if forced, at least in part, to ascend.
(2) Convective cause: Differential heating takes place over a surface, land or sea, giving birth to warm areas that, in turn, lead to a warm air parcel that starts to rise.
(3) Convergence cause: When two air masses meet at an obtuse angle, such as in the Doldrums (4.3), an ascent of air must take place.
(4) Cyclonic cause: This is often called frontal precipitation and may then be taken to include the convergence rain, since the interface of two air masses constitutes a front (Chapter 4). However, the difference is mainly that in frontal rain the two air masses have different physical characteristics and meet at an acute angle. The temperature differences among adjacent air masses accentuate the vertical motions. This is the precipitation associated with frontal passage in the temperate zones whereas the convergence type is mainly found in the tropics.

3.5.2 *Annual values*

Every station exhibits a variation in the amount of rain recorded each year. In general such variation is larger within the tropics and, expressed as a percentage of the average fall, gets greater as the amount decreases. For instance, some stations can receive one-third or three times their annual average while some desert stations can receive zero or 10 times the mean.

For any one station with an annual mean above 40 cm, there is about a 75 per cent chance that the annual totals show a normal distribution. This is a symmetrical bell-shaped distribution.which can be expressed completely in terms of the mean and the standard deviation, *s*, where

$$(N-1)s^2 = (x - \bar{x})^2,$$

where N = the number in the sample, and
x = the individual observation.

This knowledge enables probability values for receiving chosen amounts to be calculated once the standard deviation is available. It is not possible to give an exact relationship between the mean, $\bar{x}$, and the standard deviation, s, but a reasonable equation is

$$s = \bar{x}/7 + 5,$$

where both s and $\bar{x}$ are in cm, and $\bar{x}$ exceeds 50.

In order to obtain a reliable mean value, due to these fluctuations, it is necessary to use about 25 years of records because the standard error of the mean will then be about $(0.03\bar{x} + 1)$, or about 3 cm for a mean of 70 cm. The standard error of the mean gives a measure of the dispersion of the mean and is expressed as $s/\sqrt{N}$.

3.5.3 *Monthly values*

Monthly totals at a station are even more variable than annual values, and must be transformed if a normal distribution is to be used. Some workers have utilized the incomplete gamma-function in order to calculate probability levels.[3] Monthly variations are, in general, so large that care must be taken when interpreting the records for it will not be wise to assume that the average amount, or even close to it, will fall during the month. Again this variation is especially marked at the tropical stations.

3.5.4 *Diurnal variation*

It is clear that not only is the amount of rain important but so also is the temporal distribution of that amount; for instance, a 30-cm shower in 1 day during the month would be undesirable whereas if 1 cm were received every day or 2 cm every other day conditions might be ideal. The same applies to the diurnal pattern. Usually, it is desirable to have rainfall during the hours of darkness because it then causes less inconvenience and, more important, it has a chance to benefit the soil and the plants before evaporation begins to take place. It is unfortunate that in so many of the areas of the world where crop production must be improved the knowledge of time of rainfall is not known or is unsubstantiated by records.

In regions where convection plays a large part in triggering the rainfall it may be expected to reach a maximum during the late afternoon and evening. The folly of assuming this to be so can be seen if the reader consults the paper by Thompson[4] in which he shows that, in East Africa, whatever the time of day chosen there is some part of the territory in which this corresponds to the period of maximum rainfall.

3.5.5 *Altitude variation*

The presence of high land is usually sufficient to cause the air masses to ascend, cool and precipitate. This often means that, within a small area, there is an increase of rainfall with height. However, the rate at which this increase occurs varies greatly with the topography and the prevailing synoptic conditions. The increase does not of course continue indefinitely since the cooler air has a maximum moisture content that is small, so that at a level, which may vary from about 1 000 m in Hawaii to 3 000 m in East Africa, a maximum annual precipitation is noted. Above this level the decrease in amount is rapid.

3.5.6 *Drought*

Drought is really a matter of definition. The *Glossary of Meteorology*[5] defines it as

'a period of abnormally dry weather sufficiently prolonged for the lack of water to cause a serious hydrological unbalance (crop damage, water supply shortage, etc.) in the affected area.'

Palmer[6] of the US Weather Bureau has derived an equation for measuring the intensity of a drought but it is true to say that the equation would have to be altered for different climatic zones owing to the complicated interrelationships between rainfall and evaporation that exist when a drought is imminent or occurring. For instance, in the UK it is general practice to declare a 'drought' when there have been 15 consecutive days without rain, while if such a definition were used in, say, West Texas, a region where the vegetation is adapted to long dry periods, it would only cause amusement.

3.5.7 Snow

Snow is an extremely difficult phenomenon to measure because drifting and blowing takes place. Normally a measurement is made of the accumulated snow on the ground, a snow stake being used where this accumulation is large. The important factor of the water equivalent of the snow depends upon the snow depth and density. Snow density shows very great variation, having been measured in the range 0·004 to 0·91. It is general practice to assume a value of 0·1, but in areas of appreciable accumulation this often should be increased to about 0·5.

The permanent snow line at the equator is about 4 700 m, increasing to about 5 200 m in the dry air regions around the tropics and then decreasing to about 3 000 m at 45 °N and 1 400 m at 60 °N. In the southern hemisphere these values are lower. It must be noted that the permanent snow line height depends also on the direction of exposure of the slope.

3.6 Humidity

3.6.1 Introduction

The amount of moisture in the air is clearly of great importance to all water processes. The maximum amount of water that can be held by a parcel of air increases with its temperature and because of this fact there are many ways of expressing this parameter. As has been noted earlier, humidity is not an easy element to measure, especially within small volumes. Because of the dependence upon temperature the desert regions can, and do, have a higher vapour pressure than the moist polar regions.

3.6.2 Parameters

It is not the intention of this section to discuss the merits and demerits of the various parameters used for humidity specifications. But for reference purposes the commonly used ones and their definitions are given below.

Vapour pressure: The pressure exerted by the water vapour in the air, represented by e; if the air is saturated at the same temperature the term is written e_s.

Saturation deficit: The difference between e_s and e.

Relative humidity: $100e/e_s$.

Dew point: Since e_s is a function of temperature a value, T_d, can be found at which the corresponding saturation vapour pressure is equal to e; T_d is then the dew point.

Absolute humidity: The density of the water vapour expressed as mass per unit volume of air, generally g/m^3.

Specific humidity: Mass of water vapour per unit air mass.

Mixing ratio: Mass of water vapour per unit mass of dry air.

3.6.3 The dew point

During the course of the day the relative humidity normally shows a variation. However, unless an advective change of air occurs the dew point temperature is a much more conservative element. This fact makes it desirable to use the dew point for applied studies, unless the medium under observation is responsive to some other parameter. A knowledge of the dew point enables the other humidity factors to be calculated easily, once the temperature variation is also known. Because of this, it is often possible to use a constant dew point for each day since the change is relatively small compared with the other humidity parameters.

Also, it is possible, knowing the sea level conditions, to use theory to predict a lapse rate of dew point, a fact that can be extremely useful. If the relative humidity alone is known then no such technique can be used.

3.7 Air Movement

3.7.1 Introduction

From the point of view of climatology, air movement is of greater importance than its causal factor pressure. The manifestation of air flow can be thought of as being in two distinct groups, the primary flow, usually called the general circulation of the atmosphere, and the secondary flow, the local winds. The aspects of the general circulation will be discussed in Chapter 4 so that here we will consider just the secondary flow and patterns close to the surface.

on overcast days. The land breeze is the opposite phenomenon, occurring at night when the land cools more rapidly than the sea and flow from land to sea takes place. The land breeze is generally more shallow and slower than the sea breeze.

Foehn winds occur in nearly all mountain areas on the lee side of the range. They blow because ascent of air on the windward side causes a drying

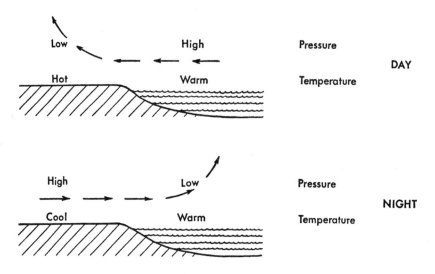

Fig. 3.6 Land and sea breezes

The highest air speeds recorded at the earth's surface are in excess of 300 km/hour but a wind has only to exceed 50 km/hour to be considered a gale, and over 115 km/hour constitutes a hurricane-force wind.

3.7.2 Secondary flow

The main types of secondary winds are land and sea breezes, foehn or chinook winds, gravity winds and whirlwinds or rotating winds.

Sea breezes occur when heating causes a higher temperature, and a lower pressure, to develop over land than over the adjacent sea. Cooler air from over the water then flows in to take the place of the ascending hot air and a cooling breeze results (Fig. 3.6). If the land region is forested the temperature differential may take many hours to reach a sufficient threshold and the sea breeze may not develop until midday, if at all. The breeze is usually relatively shallow and not of great speed. As it is also a function of the radiation it may never develop

out of the air which then descends on the lee side under dry adiabatic heating (Fig. 3.7). The foehn wind hits a region almost like a front and rapid increases in temperature are the rule, the record being about 27 °C in 2 minutes at Spearfish, S. Dakota.

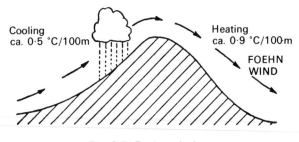

Fig. 3.7 Foehn wind

A gravity wind occurs when surface cooling at night gives rise to a difference in air density along a slope. The heavier air at the top then drains down under gravity effects and a gentle, cold wind results.

If for any reason, such as a fence, orchard or house, this flow is dammed, a cold pool or frost hollow is formed (Fig. 3.8). Extremely low temperatures can result in frost hollows: for instance, in a site in Austria, elevation 1 300 m, temperatures as low as −51 °C have been observed.[7] On a large scale they are often referred to as fall winds. These pre-suppose, in general, a high elevation adjacent to a lower level, warm surface.

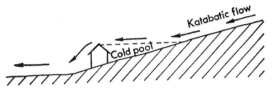

Fig. 3.8 Gravity flow

Rotating winds include dust devils, whirlwinds and sea spouts. The tornado, the most destructive of the group, usually hangs from a cumulonimbus cloud and is several hundred yards in diameter with wind speeds that on occasions have been estimated at over 500 km/hour. Over water it is called a water or sea spout. Dust devils or dust whirls are formed by strong convective activity and are characteristic of tropical deserts. They are normally only a few metres or tens of metres in diameter at the ground but can be rotating with sufficient intensity to carry small rocks around in them. The dust whirls of the cities are usually formed by the confluence of winds along streets.

3.7.3 *Variation with altitude*

The variation of wind speed with height has been studied both in the laboratory and the field. Perhaps the most useful equation for applied purposes is the differential expression[8]

$$\frac{\mathrm{d}u}{\mathrm{d}z} = az^{-\beta}$$

where z is the height above the surface, u is the horizontal wind speed, a is a constant and $\beta > 1$ for unstable conditions and <1 for stable conditions. Conditions are unstable when the temperature lapse rate is greater than the dry adiabatic rate. Values of β vary from about 1·2 to 0·75. From this equation, by integration,

$$u = az_0{}^{1-\beta}\{(z/z_0)^{1-\beta} - 1\}/(1 - \beta)$$

where $u = 0$ at $z = z_0$, the roughness height. Thus,

it is assumed that the wind speed is reduced to zero at some distance z_0 above the surface. This value of z_0 is a function of the type of surface, varying from about 0·05 mm over smooth snow, to 6 mm for closely-cropped grass to 7 cm for vegetation about 75 cm high. For tall vegetation the value of z_0 can vary since, with high wind speeds, bending takes place and the roughness height is reduced.

3.8 Evaporation and Evapotranspiration

The two parameters of evaporation and evapotranspiration are usually considered together in spite of the fact that they are two distinctly different elements of the climate.

Meteorology, and therefore climatology, concerns itself with the measurement of factors after a certain degree of standardization has been effected. This standardization is absolutely necessary so that the comparisons that play such a basic part in climatology can be made among data obtained from all parts of the world. The measurement of the evaporative power of the air should concern itself first with the study of evaporation from a standard surface. The most easily available and acceptable surface is water. Thus, it is logical to consider the measurement of evaporation from an open water surface to be a purely climatological element. Evapotranspiration is a very different element, depending as it does so much upon the type of vegetation from which the evaporation and transpiration is taking place. The problem of this element is one that must be tackled by a team of scientists and does not, strictly, deserve a place in a discussion of climatology. However, this is a book on applied climatology and, in this realm, evapotranspiration is a definite challenge. For this reason evapotranspiration is studied here.

3.8.1 *Evaporation*

Evaporation evaluation may be reached by two methods. The first, on which many investigators have worked, is by calculation. The foremost of these will be detailed later. Secondly there is observation. Under this heading things have not been so easy and this is where there exists the surprising lack of standardization. There is more agreement among specialists concerning the correctness of some of the present equations than there is with regard to the instrumentation. The calculations entail certain assumptions and need information

about an appreciable number of climatic parameters and this is never an ideal approach. It is
comparable to the approach of the scientific observer and theoretician who sets up a complex radiation energy equation, which includes assumptions
about reflectance, absorption, soil and air physical
properties and many others, simply to obtain an
estimate of the air temperature when he could
have measured it using a reliable thermometer.

For the calculation of evaporation the generally
used equation, from Penman,[9] is

$$E_0 = \frac{\Delta H + 0.5 E_a}{\Delta + 0.5} \text{ mm/day}$$

where Δ = slope of vapour pressure curve for water
　　　　　at the mean air temperature,
　　　　　T_a (mmHg/°C);
　　H = (incoming short-wave radiation) x
　　　　　(absorption coefficient) −
　　　　　(outgoing long-wave radiation);
　　E_a = 0.35 (mean saturation deficit) x
　　　　　$(1 + 0.006u)$ mm/day;
　　u　= average wind flow at 2 m, in km/day.

For details of derivation and application the
reader is referred to works by Penman.

3.8.2 Evapotranspiration

In practical problems that do not deal with an
open water surface there is need to arrive at a value
of the evapotranspiration. Like evaporation this
value has been both calculated and measured.

Penman's calculations for converting E_0, the
open water evaporation, to E_t, the potential evapotranspiration from a green vegetation never short
of water, involve multiplying by a factor f, depending upon leaf structure and daylight length. For the
British Isles this value of f ranges from 0.6 in winter
to 0.8 in summer.

Measurements of evapotranspiration have been
made using lysimeters, or large drums containing
soil and vegetation of the same type as the surrounds,
which are weighed by some very accurate balance.
Some of the errors that may arise are those due to
the lack of capillary water flow from the ground
water, edge effects and the disturbance to vegetation and soil. Most results indicate that, with
sufficient water, the value of E_t is about $0.7E_0$ but
that as the soil dries the evapotranspiration becomes
a decreasing fraction of E_0.

Chapter 4

OCEAN CURRENTS AND AIR MASSES

4.1 Introduction

This chapter is not intended to be a comprehensive discussion of air mass analysis but is intended to give a background of two main factors in climatology: ocean currents and air masses. These phenomena are manifestations of an astronomical and geological nature for, in their primary circulation, they are functions of the geometry of the solar system and the land-sea configuration. Of course, their perturbations and changes, the secondary circulation factors, are of fundamental importance to weather phenomena but it is the long-period patterns that must be understood in order to study and, perhaps, explain certain of the climatic phenomena.

4.2 Ocean Currents

The main factors that give rise to ocean currents are

(1) rotation of the earth;
(2) the velocity (speed and direction) of the prevailing winds;
(3) variations in the water density due to differences in the temperature and salinity.

The third factor is influenced by the radiation pattern that exists over the oceans season by season, in a manner similar to that in which the radiation patterns over the earth's surface influence the pressure distributions and, therefore, the general atmospheric circulation.

The ocean currents are perforce influenced by the land-sea configuration, their direction also being subject to the coriolis force which tends to impart a clockwise rotation in the northern

hemisphere and a counter-clockwise rotation in the southern. As can be seen from Fig. 4.1 the general tendency is for cold water to be carried equatorwards along the eastern margins of the oceans (western edges of the continents) and for warm water to be moved polewards along the western margins of the oceans (eastern edges of the continents).

Along the western edges of some continents the prevailing winds blow seawards. These winds cause a frictional drag that tends to move the warmer upper layers of water away from the shore faster than the lower layers causing the cold water below to upwell to the surface. Because of the unusually low sea surface temperatures for the latitude these regions (Peru, south California, S.W. Africa) are subject to frequent fogs.

Naturally, the presence of a cold current flowing along the edge of a continent tends to cool the coastal region whereas a warm current tends to warm the region, an effect that can be seen clearly in Fig. 4.1.

4.3 General Circulation

For this section the term 'general circulation' is taken to be the mean or time-averaged hemispheric circulation of the atmosphere and, for the purposes of this book it is necessary to give only a broad description, while for much fuller detail the reader is referred to the cited publications of Brunt[1] or Byers.[2]

Because of the distribution of pressure both at the earth's surface and in the upper atmosphere a basic system of planetary winds obtains. We have seen how pressure patterns vary with season, land and sea configurations and such-like,

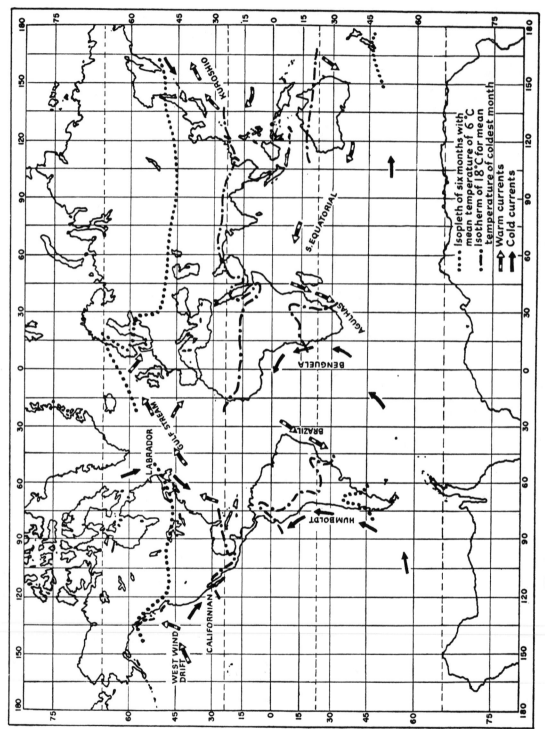

Fig. 4.1 Ocean currents and their effects on temperature

so that we cannot expect a simple circulation of winds to occur. But some general observations may be made.

At the equator, due to the intense heating of the zenith sun, the air becomes warmed, a low pressure develops and the air tends to rise. This air loses its moisture, cools in the upper atmosphere and then begins to subside, at about the tropics. By now, it is very dry and hot, helping to bring about the large desert areas of the globe. The high pressures occurring around the tropics are broken up into cells, and from these the air flows towards the equatorial low or trough giving rise to the trade winds. The air flowing from these subtropical highs towards the poles is diverted by the coriolis effect and becomes the belt of 'westerlies'. The subpolar lows, situated at about 60° latitude give rise to polar easterlies and, where these meet with the prevailing westerlies, a front (see 4.5) is formed. At the poles a region of subsiding cold air is instrumental in forming the polar high. The patterns of the general circulation in both the vertical and horizontal are shown in Fig. 4.2.

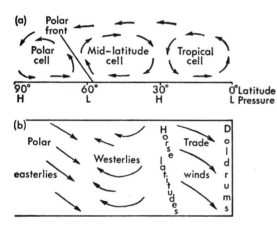

Fig. 4.2 Simplified general circulation in the vertical and horizontal

The above description refers to the average pattern of the winds and pressure but, due to the apparent migration of the sun, the regions of heating and cooling move also, giving a general tendency for the pressure and wind belts to move a few degrees northwards in the northern summer and southwards in the winter. However, the patterns of land and sea also complicate matters,

resulting in such aspects as the heat equator being a few degrees north latitude instead of at the true geographical equator.

4.4 Air Masses

The *Glossary of Meteorology*[3] gives the following definition of an air mass:

> a widespread body of air, the properties of which can be identified as
> (*a*) having been established while that air was situated over a particular region of the earth's surface, and
> (*b*) undergoing specific modifications while in transit away from the source region.
> An air mass is often approximately homogeneous in its horizontal extent, particularly with reference to temperature and moisture distribution.'

Because of the need for the source area to be large in extent, such source areas are limited around the globe. Fig. 4.3 gives an idea of where they are to be found.

These source regions give rise to the following types of air mass. It should be noted that during winter Arctic air also forms over the interiors of Canada and the Soviet Union.

Arctic (or antarctic): Cold aloft and extending to great heights; most developed in winter over ice and snow surfaces.
Polar: Often developed within subpolar highs;
  Continental: Low surface temperature, low moisture content and great stability in lower layers.
  Maritime: Initially like continental but in passing over warmer water it becomes unstable with a higher moisture content.
Tropical: Characteristics developed in low latitudes;
  Continental: Hot and very dry, being produced over subtropical arid regions.
  Maritime: Very warm and humid, being produced over tropical and subtropical seas.
Equatorial: A term sometimes used to distinguish tropical air that has stagnated in the doldrums zone of the equator.

Because of the definite characteristics shown by each air mass its advent at a certain area brings with it certain characteristics of the climate so that the frequency of occurrence of each of these

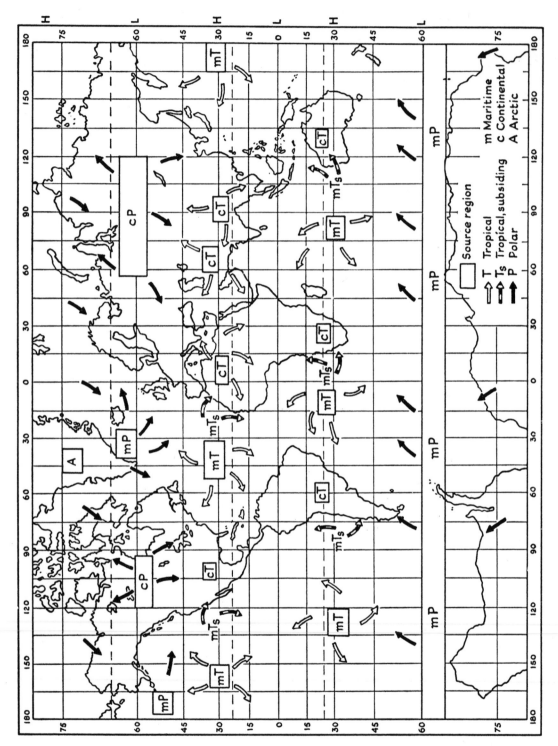

Fig. 4.3 Air masses and source regions

air masses will generate the overall picture of the climate. Of course, due to modifications these characteristics can become so altered that to state, for example, that a region experiences 25 per cent continental polar air, 35 per cent maritime polar air and 40 per cent tropical maritime air does not necessarily give a great deal of information concerning the climate of the region. However, knowledge of the source regions and the average movements of the resulting air masses is essential to an understanding of world climatology.

It can be seen from Fig. 4.3 that, because of the land-sea distribution, maritime tropical air is more frequent than the continental tropical air in both hemispheres, while continental polar occurs more frequently than maritime polar air in the northern hemisphere.

4.5 Fronts

A front can be defined as the interface or transition zone between two air masses of different density. This definition is sufficient for applied climatological purposes except that the change in density is not, by itself, important. It is the manifestations that give rise to this density change that are important; for example, temperature difference and humidity variation.

The term front, in common usage, usually implies the interface between polar and tropical air, but it may be used correctly to describe the interface between any two air masses. The front, which may start as being very well defined, will often become gradually more hazy as time progresses because of the modification that takes place to both air masses. Such is not always the case however because occasionally an outburst of air tends to reinforce one or other air mass.

Due to the change in air mass associated with the passage of a front there is usually some form of precipitation noticed at this time.

Different types of fronts are listed below.

(1) Arctic front: A semi-continuous and semi-permanent front between the arctic and the polar air; called antarctic front in the southern hemisphere.

(2) Cold front: Any front that moves so that at the surface the colder air replaces the warmer air.

(3) Occluded front: A composite of two fronts, formed as a cold front overtakes a warm front.

(4) Polar front: The semi-permanent, semi-continuous front between the polar and tropical air.

(5) Warm front: Any front that moves so that at the surface the warmer air replaces the cooler air.

A special type of front is the intertropical front, called also the equatorial trough or intertropical convergence zone. This is the convergence, or confluence, zone between the trade winds of the two hemispheres and, although often termed a front, exhibits none of the usual frontal characteristics. The major characteristics of the equatorial trough generally are a change in wind direction and an appearance of rain associated with the converging, usually moist, air masses.

4.6 Mesoscale Disturbances

The mesoscale disturbances that occur, either systematically or randomly, to the general circulation can give rise to some of the severest weather phenomena of a region. Such occurrences as the Sumatras of S.E. Asia, the winter depressions passing through the Mediterranean, the Bay of Bengal cyclones, the typhoons of the western Pacific and the West Indies hurricanes, among others, each bring certain characteristics to the climate of that area. Although their time or frequency of occurrence cannot be forecast accurately at the present time, their average occurrence and effect are known in most cases. These effects include such phenomena as gales, excessive and intense rain, storms, surges, etc. This enables climatology to take cognizance of these phenomena without being in possession of all the meteorological aspects associated with them.

Chapter 5

THE CLIMATE OF THE STANDARD CONTINENT

5.1 Introduction

It has become obvious from the previous chapters that there are many factors, simple and complex, playing important parts in forming the climate of the earth's regions. These factors may be considered, basically, to be pressure and wind patterns, global position and ocean currents.

Thus, if for the present the effect of topography is ignored, it is possible to consider a standard, or ideal, sea-level continent, stretching from the equator to the pole, and to discuss the climatic zonation that is likely to occur on such a continent. This approach helps to present a picture of the fundamental zonations that may then be compared with the actually-occurring zones on the various continental masses.

5.2 The Climatic Factors Affecting the Standard Continent

It is convenient and realistic to represent the standard continent by a shape in which, at each latitude, the width is proportional to the global land area at that latitude. Hence the shape of the next figures.

In Fig. 5.1 are given the main ocean currents that would affect such a continent and the mean annual pressure and wind patterns that would prevail.

These show that the eastern seaboard would be warmed in the lower latitudes by ocean currents while on the west margin, above about latitude 40°, a similar effect would be felt. At lower latitudes on the west, however, there would be a distinct cooling due to the movement of cold water equatorwards from higher latitudes.

The semi-permanent subtropical highs and subpolar lows will be approximately as shown, and these will give rise to the air flow indicated.

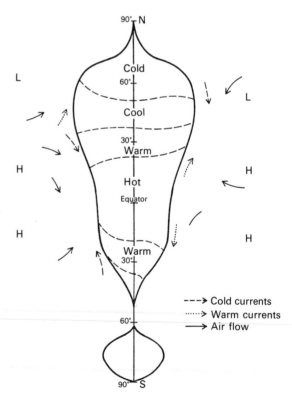

Fig. 5.1 Ocean currents, pressure patterns and thermal zones for the standard continent

There will, of course, tend to be a high-pressure region over the northern continent in the winter and a low pressure over the tropical region during the summer.

5.3 Temperature Patterns on the Standard Continent

Considered on an annual basis, the main factors contributing to the thermal patterns are as follows.

(1) The ocean currents.
(2) Warm air masses from the oceans and cool air masses from the continental interiors in winter; cool air masses from the oceans and warm air masses from the continents in summer.
(3) Latitude, in its effect on the radiation climate.

These factors will then lead to an annual thermal pattern on the standard continent roughly as shown in Fig. 5.1. The summer and winter patterns are given in Figs 5.2 and 5.3.

5.4 Precipitation Patterns on the Standard Continent

Owing to the pronounced change that occurs in precipitation patterns between the summer and the winter seasons it is necessary to consider these separately.

During the winter season (December–February) the general pattern is approximately as shown in the northern hemisphere of Fig. 5.4. The polar fronts, along which rainfall is occurring may, of course, change position radically from day to day but their sphere of influence is indicated by the shaded area of cyclonic or frontal rain. The front is not shown as continuous across the continent because the cold high tends to block the movement of many rain-bringing cyclones moving from the west. Precipitation naturally occurs in the continental interior, but in the form of snow, this being equivalent to a small amount of rain. Along the equator the convergence belt of the trade wind is still sufficiently close to show an associated

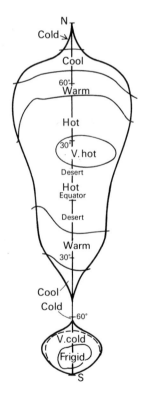

Fig. 5.2 Temperature patterns in summer (northern hemisphere)

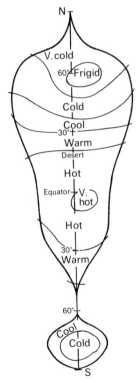

Fig. 5.3 Temperature patterns in winter (northern hemisphere)

rain belt, especially when there is sufficient heating to induce convective occurrence as well as convergence. Along the eastern sea-board around the 40° latitude there is often some rainfall associated with convective showers over the relatively warm ocean that have been blown landwards by the outflow from the subtropical high.

During the summer (June–August) the equatorial convergence zone has moved into the northern hemisphere and most of the tropical belt experiences rain associated with this manifestation. The polar front has now retreated towards the 60°–70° latitude and gives rise to small amounts of rainfall around this area. In the interior of the continent, in regions where there is a source of moist air, the intense convection leads to heavy showers. Along the eastern margin of the continent there is general rainfall due to the presence of moist onshore winds that are subject to convective effects. The summer pattern can be seen in the southern hemisphere in Fig. 5.4.

Although the consideration of the summer and the winter patterns gives a good idea of the annual zonation, there are two special regions that must be considered separately.

During the winter the cyclones moving from the west are unable to penetrate far into the continent because of the blocking effect of the high pressure. However, as spring approaches the high weakens and some rain-bearing winds can break through into the interior. Thus, this normally dry region has a tendency for a spring maximum of rainfall.

In high latitudes along the western edge of the continent and during the autumn as the land begins to cool off, the relatively warm moist air moving in from the oceans tends to give a seasonal maximum.

With the patterns obtaining during the four seasons it is now possible to make a composite picture of the rainfall regimes. This is given in Fig. 5.5. It is interesting to note that the areas

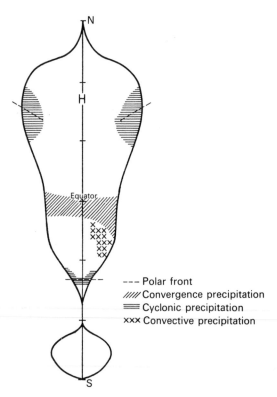

--- Polar front
///// Convergence precipitation
=== Cyclonic precipitation
xxx Convective precipitation

Fig. 5.4 Rainfall patterns in summer (southern hemisphere)

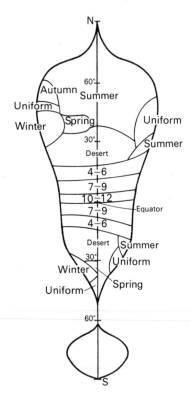

Fig. 5.5 Annual rainfall patterns

of uniform rainfall are quite small and that the desert region extends over a wide latitude band. In the region near the equator the approximate number of wet months has been given as there is little meaning to the terms spring, summer, autumn and winter, in this area (Chapter 6).

5.5 Effect of Shape and Topography

There are innumerable possible variations in the shape of a continent from that of the assumed triangle but, to indicate the type of climatic patterns that may result from deviations from 'standard' conditions, it is sufficient just to consider two extreme variations.

(1) If the width of the continent in the higher latitudes is very great, for example, North America and Asia, there will be a large source area for cold air in winter. This will then lead to extreme winter cold and intense heat, for that latitude, during the summer, because of the radiation absorption plus the lack of moist air so far inland.

(2) If the width of the continent around the tropical regions is small, as, for example, in North America, there will be little continental tropical air and a smaller desert area will result.

Topography, in the form of mountain ranges, will show an effect on the climate dependent upon the orientation of these ranges. For example, if the mountains run N–S the tropical air of summer can sweep right into the continental interior while, during the winter, the cold arctic or polar air can move down towards the equator. This does occur in America where the Rockies and the Andes are both N–S mountain chains. Warm moist air penetrates as far as Canada in summer while the cold 'Northers' reach as far south as Mexico during the winter.

If the ranges run E–W they are effective in blocking great latitudinal movements of both tropical and polar air masses. Thus the interior can get very cold in winter, since no outflowing occurs, while the tropical region can experience a very hot season as the warm air is held within bounds. This pattern occurs in Asia where the Tibetan plateau and the Anatolian–Persian ranges run E–W.

Chapter 6

CLIMATIC CLASSIFICATIONS

6.1 Introduction

There are no two places in the world that experience exactly the same climate, but it is possible to identify areas with similar climates. This 'grouping' or 'analogue' method is referred to as climate classification. Normally a classification is developed to assist in some special problem, such as a study of interrelationships between climate and vegetation. There is no possibility of deriving a classification that is completely efficient for all applications but any system developed should be easy to apply, be based on meteorological principles, collate the data into a manageable form and be directed towards limited, well defined objectives.

Because of the large number of reporting stations for which measurements are obtainable, the basic elements of temperature and precipitation are the ones most often used – although it would be preferable if world-wide, accurate data were available to base a classification on the fundamental parameters of heat balance and water balance. In most classifications the resulting zones are given everyday terms (such as hot or temperate) or constructed ones (such as mesothermal) but the classes have to be based on the use of numerical thresholds. These thresholds are not to be ascribed magical properties: they are used to present the broad picture.

6.2 Classification Details

So as to have a ready appreciation of the various important climatic zones, a simple classification is adopted here that uses the threshold values given in Table 6.1. With this system 32·4 per cent of the world's land area is in zone A, 23·3 per cent in B, 16·2 per cent in C, 14·8 per cent in D and 13·3 per cent in E. Deserts comprise about 21 per cent of the land area and zones with a summertime maximum of rainfall some 40 per cent. Due to altitude 11 per cent of the land area is put into a sub-zone, H, the highland. The biggest zones are the long winter, summer rain (DS), with 13·3 per cent and the cold, summer rain (ES) with 13·1 per cent. Together with the warm desert (BF) and hot, short drought (A2) the four zones cover nearly half the land surface. This climatic zonation is shown in Fig. 6.1.

6.3 The Tropics (A Zone)

In a belt set astride the equator, and limited in the poleward direction by deserts, lies the tropical zone. A characteristic of this zone is the small seasonal variation in the incoming solar energy, a fact that makes for relative uniformity in temperature and which also means that the driving force for atmospheric variations is not subject to rapid changes. In some parts the temperature criterion is not met due to altitude but when the temperatures are reduced to sea-level values, using a correction of 5·6 °C per 1 000 m, they do fall in the A zone. These are referred to as highland tropics, HA, the only region in the A zone where artificial heating is regularly needed at some season. There are three distinct tropical land regions:

(a) Africa and southern Arabia;
(b) Central and South America;
(c) India, S.E. Asia, northern Australia and the Pacific Islands.

6.3.1 *Hot, wet (A1)*

In this zone, which has no more than two dry

TABLE 6.1

Classification

Temperature

Hot (A)	mean temperature of all months	$\geqslant 18\,°C$
Warm (B)	mean temperature of all months	$\geqslant 6\,°C$
Short winter (C)	mean temperature of 7–11 months	$\geqslant 6\,°C$
Long winter (D)	mean temperature of 3–6 months	$\geqslant 6\,°C$
Cold (E)	mean temperature of 0–2 months	$\geqslant 6\,°C$
Highland (H)	when altitude places station in a different thermal zone from what it would be if at sea level	

Rainfall

Desert (F) $R < (16 + 0.9T)$ (R in cm, T in °C)

In A zone 1	10–12 months each with $\geqslant 50$ cm	
2	7–9 months each with $\geqslant 50$ cm	
3	3–6 months each with $\geqslant 50$ cm	

In A, B, C, D, E zones   Uniform (U): $\dfrac{\text{rainfall in driest consecutive 3 months}}{\text{rainfall in wettest consecutive 3 months}} > 0.5$

spring (V)	central month of wettest quarter, in March, April or May
summer (S)	central month of wettest quarter, in June, July or August
autumn (A)	central month of wettest quarter, in September, October or November
winter (W)	central month of wettest quarter, in December, January or February
'	wettest quarter has 40–60 per cent of the annual total
"	wettest quarter has over 60 per cent of the annual total
continental	annual temperature range $> 17\,°C$
extreme continental	annual temperature range $> 34\,°C$

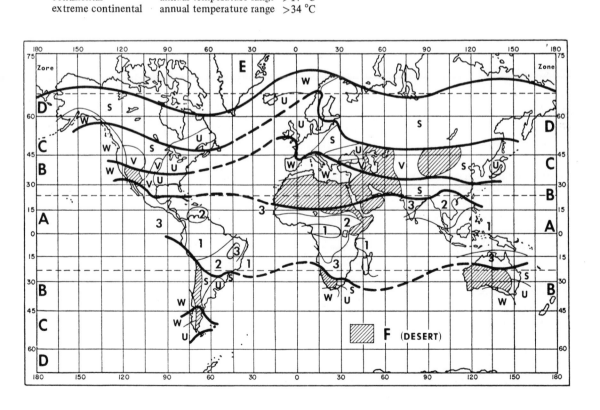

months, and almost encircles the globe, the I.T.C.Z. — that zone of low pressure at the meeting of the Trade Winds — plays a major role, but it can be supplemented by on-shore winds or monsoon effects. The obvious break in the pattern is in eastern Africa where the highlands generate their own special circulation. This is the region where there is the combination of high temperatures and sufficient soil moisture for prolific vegetation growth. This is the zone of the tropical rain forest. Temperatures are uniformly high during every month, although some stations experience diurnal relief due to their continental location while the island or coastal situations get no such alleviation of the enervating conditions. However, even inland, the high humidity makes for a smaller diurnal variation than normal and the incoming radiation load is reduced due to absorption by water vapour.

6.3.2 *Hot, short dry spell (A2)*

On the poleward edges of the hot, wet zone (A1) exists a band of land where a relatively dry spell of from 3 to 5 months occurs. The regions lie outside of the continuous influence of the I.T.C.Z. and so experience a season of hot, dry conditions. The vegetation generally changes from the forest on one edge to a deciduous type woodland on the poleward side. It is an area in which most agricultural development can take place for moisture is sufficient and reliable, yet it is not too excessive.

6.3.3 *Hot, long, dry spell (A3)*

This region is normally on the equatorward fringe of the major desert regions and experiences some 6–9 months of low rainfall. An exception is found in north-east Brazil where local topographic and atmospheric peculiarities give rise to an A3 region, called the 'Sertao'. It must be realized that the expression 'low' rainfall is to be judged in the light of the great evaporation that the combination of high temperatures and low humidities (during the dry season) can cause. The vegetation here is normally savanna and scrubland. It is a region particularly prone to suffer from the normal and expected vagaries of the weather, for a small deviation of the 'drier' side of the average can bring disasters, witness the south Saharan tragedies of the 1970s, the Botswanaland droughts of the 1960s

and the sporadic famines in India at the 'failure of the monsoon'.

6.4 The Subtropics (B Zone)

Here, in a wide belt to the poleward sides of the A zone, is a zone in which mean monthly temperatures are always warm, 6 °C or greater, yet they do have a 'cooler' season. As with the Tropics there are also highland subtropical (HB) areas where altitude dominates the thermal pattern. The B zone covers about a quarter of the earth's land surface but much of this (about 40 per cent) is desert and is little used, except for the mineral wealth — especially oil. The subtropical zone is found on all continents, actually most of Australia is in this zone.

6.4.1 *Warm, uniform rain (BU)*

This region has some characteristics in common with the A and the C zones for in both North and South America it extends completely across the B zone. The other two locations of BU are in South Africa and south-eastern Australia. Due to the uniformity of the rain the summers can be unpleasantly humid, particularly on the equatorward extremes. The winter period is generally very pleasant and attracts many people from the more rigorous climates, in addition to migratory birds. A feature unique to the BU region in North America is the occasional entry of very cold air masses into the subtropics, plunging temperatures well below freezing with often calamitous effects upon the vegetation. These 'northers' are much more intense than the corresponding 'pamperos' in Argentina, due to the larger land area from which they are derived.

6.4.2 *Warm, summer rain (BS)*

This region is found on all the continents save Europe and always adjoins either an A2 or an A3 region. In much of the BS region about half the annual rainfall total is recorded during three warmest months, an amount that increases to 65–70 per cent in parts of India, where the monsoon circulation dominates. The combination of high temperatures and rainfall can lead to uncomfortable summers but winters are normally most pleasant (note that the remarks under BU apply here also).

The hottest period in the 'monsoon' region is generally immediately before the onset of the rains.

6.4.3 *Warm, winter rain (BW)*

This is the region experiencing the conditions that are often called 'the Mediterranean climate', and said by many to be the most desirable for man, with its 'warm, wet winters and hot, dry summers'. It is found on all the continents, but the largest region is around the Mediterranean Sea. In this latter area the northern section (S. France, Italy and Yugoslavia) exhibit a slight autumnal maximum of rainfall, due to flow off the warm sea, while central Spain has a spring, or vernal, maximum because the formation of a high pressure (cold air) over the Iberian plateau blocks some of the winter rain-bearing depressions. During the summer all parts of this BW region are generally under the influence of dry, descending air from the semi-permanent oceanic Highs. All areas can experience very hot, windy and dusty summer-time outbreaks from their adjacent large deserts, although the northern Mediterranean lands are less often affected. The natural vegetation in BW is of the xerophytic variety capable of withstanding the temperature-drought stress by possessing waxy or stiff leaves, long roots or water storage characteristics.

6.4.4 *Warm, desert (BF)*

This is an extremely large area, that, together with the relatively small tropical desert region (AF), included here for convenience, covers about one-eighth of the land surface. Because of the extreme aridity of the air the summertime temperatures can get unbearably high (50 °C and above), and here are found world-record high temperatures, sunshine amounts and low rainfalls. The only alleviation of the summer heat is due to the rapid out-flow of heat at night so that diurnal temperature fluctuations are extreme (20–25 °C is not unknown). Vegetation is sparse but occasionally, after brief showers, the desert truly does burst into blossom.

6.5 The Short-Winter Zone (C)

Poleward of the subtropics occurs a zone in which there is a short period, 1–5 months, during which the mean monthly temperature falls below 6 °C. This C zone comprises about one-sixth of the earth's land areas, almost all of which is found in the northern hemisphere, only Chile and Argentina extending sufficiently poleward in the southern hemisphere. The zone is centred along the 45th latitude but warm ocean currents take the boundary within the Arctic Circle to Norway but, where the influence of the bitter cold air from Siberia is felt, the limit comes down to 30°N. The zone exhibits many varieties of rainfall patterns. Vegetation is of the deciduous type and crops can benefit from the long summers.

6.5.1 *Short winter, uniform rain (CU)*

This region is to be found in five areas: (1) north-east USA and south-east Canada; (2) north-west Europe; (3) north of the Black Sea; (4) northern half of Honshu, Japan; (5) southern tip of Chile. All of these are in the belt of prevailing westerly winds, so that areas (1) and (4), which are on the eastern side of large land masses, can experience much greater temperature extremes (hotter summer days and colder winter days) than areas (2) and (5), which are situated on the western side of their continent. The Black Sea area occupies a thermal position about mid-way between the two types.

6.5.2 *Short winter, spring rain (CV)*

This region has a spring maximum of rainfall caused by an interesting set of meteorological conditions. During the winter the cold, dense air masses located over the interiors of North America and Asia tend to block the penetration of the rain-bearing depressions but during springtime, when the highs weaken and retreat northward, the area can benefit from these frontal phenomena. Such rain comes at a most opportune time for crop planting and the grazing herds. Three of the areas, north-eastern Turkey, Uzbek/Kazakh and the central Rockies of the USA, have much of their land actually classified as DV(HCV), due to their elevation.

6.5.3 *Short winter, summer rain (CS)*

The three sections of this region are to be found in the centre of the North American continent, in eastern Europe and in the north-east China/Korea area. The rainfall is associated with the inflowing air caused by the formation of the summertime low pressure (warm, less dense air) over the continents. In the European section the rainfall is less concentrated than in North America, where about half the annual total falls during a three-month period, and in eastern Asia where the fraction may rise to three-quarters on the northern edge.

6.5.4 *Short winter, winter rain (CW)*

This is a single, large region, except for very small areas in south central Chile and north western Turkey. Along the west coast of North America, from the Alaskan panhandle to Oregon, there is a dominant, subsiding air flow off the North Pacific High during the summer and rain is infrequent. However, during the autumn and early winter the cyclonic rain (from the southward moving frontal zone) is reinforced by the movement on-shore of convective showers triggered by the sea, at this time warmer than the land. The CW climates are of the maritime type, with relatively small daily and annual temperature ranges.

6.5.5 *Short winter, desert (CF)*

This region is found in two large areas, around the Caspian Sea and in the Sinkiang Province of China. Both areas are remote from moisture sources and experience truly continental conditions.

6.6 The Long-winter Zone (D)

This zone, covering about one-seventh of the land, is almost exclusively within the northern hemisphere, only around Cape Horn does it appear south of the Equator. In the D zone summers are short, afternoon temperatures averaging generally only around 15–20 °C, and long, dark winters are the rule.

6.6.1 *Long winter, uniform rain (DU)*

An unusual region where annual rainfall amounts are quite large, 500–1 500 mm, and are distributed fairly uniformly throughout the year. The only location, save for a tiny part of south-east Iceland, is in eastern Canada. Summertime rain occurs as in the zone DS but the meeting of air masses off the warm Atlantic waters and those off the cold Hudson Bay and Davis Strait (Labrador Current) yield plenty of winter precipitation also.

6.6.2 *Long winter, summer rain (DS)*

This, the largest region, covering more than one-eighth of the land area, has two subdivisions, one with continental and one with extremely continental conditions. The extremely continental subdivision is, naturally, to the eastern side of the continents. The summertime rain, while not great in amount due to the coolness of the air, generally reaches its maximum in late summer when the marshy, mosquito-ridden sections have thawed and can act as an almost continuous water surface.

6.6.3 *Long winter, winter rain (DW)*

Two small areas, northern Iceland plus Norway, and southern Alaska, are the only examples of DW climates. These areas are kept warm enough in winter so that most of the precipitation is in the form of rain and sleet, not snow. The temperature patterns are maritime, in sharp contrast to those in DS. Rainfall amounts in southern Alaska are much greater than in the European area.

6.7 Cold Zone (E)

This zone, over 13 per cent of the world's land area, is almost all in the ES region and is very similar to the DS region, except that summertime temperatures are now, at maximum, about 10–13 °C. Due to the cold temperatures precipitation amounts are very low.

Perhaps some mention should be made of those areas, included in E zone, where the mean temperatures never go above freezing. Such areas can occur all the way from the Equator (Kilimanjaro and Chimborazo) to the Poles, with Vostok the example *par excellence*, for its warmest month averages −33 °C.

6.8 Special Climates

6.8.1 *Mountain climates*

Due to the effect of altitude, mountains cause climates different from those of the adjacent lowlands. These mountain climates may differ not only with altitude but also on different sides of the mountain. A direct influence due to altitude is noticed by a decrease in pressure and temperature. In Spanish-speaking countries the *tierra caliente* (hot lowland) gives way to the *tierra templada* (temperate zone) and finally to the *tierra fria* (cold uplands); in Ethiopia the belts are Kolla, Woina Dega and Dega. This zonation also infers changes in the vegetation. Generally, the mountain climate is characterized by decreased absolute humidity and an increase in solar radiation.

Precipitation, while normally showing an increase with height to a maximum belt, generally shows the same seasonal distribution as on the lowlands. Some exceptions do occur when upper-air converging flow brings rain to the slopes and not the lowland, for example, on Mt. Kenya. Rain shadow effects are often noticeable on mountains in the trade-wind latitudes; for instance in Hawaii totals can fall from 10 000 mm/year to 400 mm/year in a distance of less than 20 km.

The effect of large mountain ranges is to prevent some air-mass movements from taking place. The Rocky Mountains and the Andean chain act as natural barriers both to keep marine influences in a narrow band and to reduce the penetration of cold (or warm) air masses to the coast. In a similar manner the Himalayan massif shuts out the extreme winter cold of Tibet and Siberia from India while preventing the moisture-laden air from the warm Indian Ocean from reaching Takla Makan and Kazakhstan.

6.8.2 *Marine climates*

These are found in regions where, either throughout the year or seasonally, the air masses are almost exclusively from over a large water surface. Because of this there is more moisture in the air and thermal fluctuations are reduced. Additionally, since the water body responds more slowly than land to the radiation changes there is a retardation in the times of maximum and minimum temperatures (Table 6.2).

TABLE 6.2

Marine vs. continental climates

Annual temperature range	M < C
Diurnal temperature range	M < C
(April temperature)/(October temperature)	M < C
Sunshine	M < C
Summer rainfall	M < C
Time lag of maximum and minimum temperature	M > C
Cloud (autumn and winter)	M > C
Rainfall	M > C
Relative humidity	M > C

6.8.3 *Continental climates*

This is almost an antithesis of the marine climate for the air masses are now almost exclusively from over a large land mass. Therefore, there is less moisture and thermal fluctuations are greater. In Table 6.2 the important differences between the marine and continental climates are detailed.

6.9 Common Classifications

There are many various climatic classifications[1] but three are especially worthy of note due to their widespread use and particular applicability. Like most classifications these have been overextended in their use and subject to much modification but each has a role of importance in climatology.

The oldest, and most generally used, is due to Köppen,[2] a German scientist who was particularly interested in the relationship between climate and vegetation. The system, derived and modified 60 to 70 years ago, uses letters to indicate various climatic characteristics – the first letter, or two letters, capitalized, relating to temperature or a temperature–rainfall pattern; the second letter, lower case, is a rainfall variable and the third another temperature aspect. A simplified version is given (Table 6.3). Köppen included many other letters to signify such items as 'monsoon' temperature pattern (g), warmest month in autumn (v), fog incidence (n), and numerous rainfall patterns. There have been suggestions recently that the arid regions should be given a thermal letter, for example, BSh could be BASh, while a polar maritime (EM) zone should be included in which the mean temperature of the coldest month is greater than −7 °C. These appear to be excellent modifications. The major problems with the Köppen classification are that it is not easy to remember all the symbols of the complete classification, and that the zones can show vast variations within themselves. For example, Af includes Entebbe with 1·5 m and Debundscha with 10·3 m of rain a year, Santos (Brazil) with an annual temperature range of 8 °C (19 °C to 27 °C) and Andagoya (Colombia) with 1 °C (27 °C to 28 °C); BSk includes Mogador (Morocco) with mean monthly temperatures varying from 14 °C to 20 °C and Ulan Bator (Mongolia) with −26 °C to 16 °C, and Cfb has regions with from 0·5 m to 2 m of rain per year.

The Thornthwaite classification,[3] used only in

TABLE 6.3

Köppen climatic classification

A	mean temperature of coldest month >18 °C
C	one or more months with mean temperature <18 °C; none <−3 °C; at least one >10 °C
D	one or more months with mean temperature <18⁰°C; one or more <−3 °C; at least one >10 °C
E	no mean monthly temperature >10 °C

(R = mean annual precipitation (cm); T = mean annual temperature (°C)).

(*a*) Winter precipitation = 70 per cent R received in six cooler months (October to March in Northern Hemisphere).
(*b*) Summer precipitation = 70 per cent R received in six warmer months
(*c*) When neither (*a*) or (*b*) applies.

If (*a*) holds and if $R > 2T$	A, C, D
$< 2T$ but $>T$	BS (steppe)
$< T$	BW (desert)
If (*b*) holds and if $R > 2(T + 14)$	A, C, D
$R < 2(T + 14)$ but $>(T + 14)$	BS
$R < (T + 14)$	BW
If (*c*) holds and if $R > 2(T + 7)$	A, C, D
$R < 2(T + 7)$ but $>(T + 7)$	BS
$R < (T + 7)$	BW

A	f	every month has at least 6 cm precipitation
	m	$R > 100 + 25$ (6-driest month)
	w	$R < 100 + 25$ (6-driest month)
B	h	$T > 18$ °C
	k	$T < 18$ °C, warmest month >18 °C
	k′	$T < 18$ °C, warmest month <18 °C
C, D	a	hot summer, warmest month > 22°C
	b	warm summer, warmest month <22 °C, 4 to 12 months >10 °C
	c	cool summer, warmest month <22°C, 1 to 3 months >10 °C
	s	summer dry, precipitation of driest month in warm season <1/3 wettest winter month and <4 cm.
	w	winter dry, precipitation of driest month in winter season <1/10 wettest summer month
	f	humid, when s and w do not apply
D	d	severe winter, coldest month < −38 °C
E	F	all monthly means below freezing
	T	not all monthly means below freezing

North America, is based on the two parameters: temperature efficiency T.E., where

$$\text{T.E.} = 0.25 \sum_{i=1}^{12} (\overline{T}_i - 32)$$

where $\overline{T}_i$ is the mean temperature for the *i*th month (°F); and the precipitation effectiveness P.E., where

$$\text{P.E.} = 115 \sum_{i=1}^{12} \{r_i/(\overline{T}_i - 10)\}^{10/9}$$

r_i is the rainfall of the *i*th month (inches). In addition, there is a factor for the seasonal pattern of the P.E. This classification was the first, after Linsser's in 1868, to investigate an evaporative term and was a bold step forward. Like Köppen's it has been over-extended in use, especially since the expressions for both T.E. and P.E. are empirical relationships that do not apply with constant accuracy to all areas of the world.

A more recent development in climatic classification is the concept of life-zone ecology developed by Holdridge.[4] It has been developed in tropical countries and is applied in land-use policy decisions. Plagiarizing the authors' lucid papers, the concept can be summarized as follows.

A life zone is defined as a uniformly weighted division of earth's climate which supports a distinct

TABLE 6.4

The life-zone classification

Latitude	Altitude	Mean Annual T_{bio} (°C)	Mean annual precipitation (mm)							
			>8000	4–8000	2–4000	1–2000	500–1000	250–500	125–250	<125
Tropical		24–30	Rain forest	Wet forest	Moist forest	Dry forest	Very dry forest	Thorn woodland	Desert scrub	Desert
Subtropical	Pre-montane	18–24		Rain forest	Wet forest	Moist forest	Dry forest	Thorn woodland	Desert scrub	Desert
Warm temperate	Lower montane	12–18		Rain forest	Wet forest	Moist forest	Dry forest	Thorn steppe	Desert scrub	Desert
Temperate	Montane	6–12			Rain forest	Wet forest	Moist forest	Steppe	Desert scrub	Desert
Boreal	Subalpine	3–6				Rain forest (Rain Paramo)	Wet forest (Paramo)	Moist forest (Puna)	Dry scrub	Desert
Subpolar	Alpine	1·5–3					Rain tundra	Wet tundra	Moist tundra	Dry tundra
Polar	Nival	<1·5								

set of plant associations, a plant association being a dominant community of plants which, in its native natural state, has a physiognomy distinct from that of all other plant associations. There are four types of these:

    (a) *Climatic association* – Plant community growing on zonal soil in zonal climate. (There is only one life zone in each climatic association.)

    (b) *Edaphic association* – Plant community growing on azonal (not sufficient time, horizon not well differentiated) or intrazonal (bedrock and relief dominate, marshes) soil.

    (c) *Atmospheric association* – Plant community growing in an azonal climate, e.g. Mediterranean climate, monsoon climate, cloud forest, mountain ridges.

    (d) *Hydric association* – Plant community growing where soil is covered with water for all, or almost all, of the year.

Zonal soils have been defined as having easily identifiable horizons, resulting from appreciable climatic and biological influence and having a definite correlation with climate; a zonal climate, however, is one in which there is a normal average climate with normal average distribution of precipitation and moisture in accordance with the annual means of biotemperature (T_{bio}) and precipitation; actually defined via the potential evapotranspiration ratio (P.E.R.). The mean annual biotemperature is actually the sum of mean hourly temperatures in °C (<0 °C and >30 °C count as zero) divided by 24 x 356, while the P.E.R. is the total amount of water that potentially could be utilized by the normal mature vegetation of a certain site (includes zonal soil in zonal climate) and is given by

P.E.R. = (T_{bio} x 58·93)/ annual precipitation, where T_{bio} is in °C, and annual precipitation is in mm. The normal average distribution of precipitation is shown below:

P.E.R.	16	8	2	1	>0.5	<0.5
Effectively dry months	12	10	6	4	2	0

where an effectively dry month occurs when soil moisture reserve is down to 50 per cent of field capacity and field capacity is taken as 10 per cent of the annual precipitation. It is difficult to represent the three dimensional picture in two dimensions but, in Table 6.4, an approximation is presented. To take an example, an area with T_{bio} = 10 °C and an annual precipitation of 1 500 mm could be

    (1) a cool temperate wet forest,
    (2) a warm temperate montane wet forest,
    (3) a subtropical montane wet forest,
or (4) a tropical montane wet forest.

The decision as to which is correct is made by increasing T_{bio} at a rate of 6 °C per 1 000 m elevation, to obtain the basal or sea-level temperature and zone. If, as an example, the area had been at 2 000 m then T_{bio} would be 10 °C + (2 x 6) ° C = 22 °C, in the subtropical belt. So the result is (3).

    The frost, or critical temperature, line is the one great anomaly. As Holdridge remarks 'on opposite sides of the critical temperate lines the taxonomic lists of plant species are markedly distinct. Cultivated plants are likewise generally different on either side of the line. The change in land use in the humid tropics where coffee is cultivated up to the top of the Premontane altitudinal belt and then gives way sharply to dairy pastures or to grain or potato cultivation is a clear-cut example of the significance of the line.'

SINGAPORE, MALAYA: 1°18′N, 103°50′E, 9 m A1U zone

	Jan	Feb	Mar	Apr	May	Jun	Jul	Aug	Sept	Oct	Nov	Dec	Average for year	No. of years
Mean maximum temperature (°C)	30·0	31·1	31·1	31·1	31·7	31·1	31·1	30·6	30·6	30·6	30·6	30·6	30·6	39
Mean minimum temperature (°C)	22·8	22·8	23·9	23·9	23·9	23·9	23·9	23·9	23·9	23·3	23·3	23·3	23·3	39
Precipitation (mm)	251	173	193	188	173	173	170	196	178	208	254	257	Total for year 2 414	64

KISANGANI, ZAIRE: 0°26′N, 25°14′E, 410 m A1 zone

	Jan	Feb	Mar	Apr	May	Jun	Jul	Aug	Sept	Oct	Nov	Dec	Average for year	No. of years
Mean maximum temperature (°C)	31·1	31·1	31·1	31·1	30·6	30·0	28·9	28·3	29·4	30·0	29·4	30·0	30·0	8
Mean minimum temperature (°C)	20·6	20·6	20·6	21·1	20·6	20·6	19·4	20·0	20·0	20·0	20·0	20·0	20·6	8
Precipitation (mm)	53	84	178	157	137	114	132	165	183	218	198	84	Total for year 1 703	14

MANAOS, BRAZIL: 3°08′S, 60°01′W, 42 m A1 zone

	Jan	Feb	Mar	Apr	May	Jun	Jul	Aug	Sept	Oct	Nov	Dec	Average for year	No. of years
Mean maximum temperature (°C)	31·1	31·1	31·1	30·6	31·1	31·1	31·7	32·8	33·3	33·3	32·8	32·2	31·7	11
Mean minimum temperature (°C)	23·9	23·9	23·9	23·9	23·9	23·9	23·9	23·9	23·9	24·4	24·4	23·9	23·9	11
Precipitation (mm)	249	231	262	221	170	84	58	38	46	107	142	203	Total for year 1 811	25

KINSHASA, ZAIRE: 4°20′S, 15°18′E, 324 m A2 zone

	Jan	Feb	Mar	Apr	May	Jun	Jul	Aug	Sept	Oct	Nov	Dec	Average for year	No. of years
Mean maximum temperature (°C)	30·6	31·1	31·7	31·7	31·1	28·9	27·2	28·9	30·6	31·1	30·6	30·0	30·0	8
Mean minimum temperature (°C)	21·1	21·7	21·7	21·7	21·7	19·4	17·8	18·4	20·0	21·1	21·7	21·1	20·6	8
Precipitation (mm)	135	145	196	196	157	8	3	3	30	119	221	142	Total for year 1 355	12

Bangkok, Thailand: 13°45'N, 100°28'E, 2 m A2 zone

	Jan	Feb	Mar	Apr	May	Jun	Jul	Aug	Sept	Oct	Nov	Dec	Average for year	No. of years
Mean maximum temperature (°C)	31·7	32·8	33·9	35·0	33·9	32·8	32·2	32·2	31·7	31·1	30·6	30·6	32·2	37
Mean minimum temperature (°C)	20·0	22·2	23·9	25·0	25·0	24·4	24·4	24·4	24·4	23·9	22·2	20·0	23·3	37
Precipitation (mm)	8	20	36	58	198	160	160	175	305	206	66	5	Total for year 1 397	44

Caracas, Venezuela: 10°30'N, 66°56'W, 1040 m A2 zone

	Jan	Feb	Mar	Apr	May	Jun	Jul	Aug	Sept	Oct	Nov	Dec	Average for year	No. of years
Mean maximum temperature (°C)	23·9	25·0	26·1	27·2	26·7	25·6	25·6	26·1	26·7	26·1	25·0	25·6	25·6	30
Mean minimum temperature (°C)	13·3	13·3	14·4	15·6	16·7	16·7	16·1	16·1	16·1	16·1	15·6	14·4	15·6	30
Precipitation (mm)	23	10	15	33	79	102	109	109	107	109	94	46	Total for year 836	46

Nairobi, Kenya: 1°16'S, 36°48'E, 1820 m HA2 zone

	Jan	Feb	Mar	Apr	May	Jun	Jul	Aug	Sept	Oct	Nov	Dec	Average for year	No. of years
Mean maximum temperature (°C)	25·0	26·1	25·0	23·9	22·2	21·1	20·6	21·1	23·9	24·4	23·3	23·3	23·3	15
Mean minimum temperature (°C)	12·2	12·8	13·9	14·4	13·3	11·7	10·6	11·1	11·1	12·8	13·3	12·8	12·8	15
Precipitation (mm)	38	64	124	410	157	46	15	23	30	53	109	86	Total for year 1 155	17

Niamey, Niger: 13°31'N, 2°06'E, 215 m A3 zone

	Jan	Feb	Mar	Apr	May	Jun	Jul	Aug	Sept	Oct	Nov	Dec	Average for year	No. of years
Mean maximum temperature (°C)	33·9	36·7	40·6	42·2	41·1	38·3	34·4	31·7	33·9	38·3	38·3	34·4	36·7	10
Mean minimum temperature (°C)	14·4	17·2	21·7	25·0	26·7	25·0	23·3	22·8	22·8	23·3	18·3	15·0	21·1	10
Precipitation (mm)	0	2	5	8	33	81	132	183	91	13	1	0	Total for year 549	10

DODOMA, TANZANIA: 6°10'S, 35°46'E, 1120 m A3 zone

	Jan	Feb	Mar	Apr	May	Jun	Jul	Aug	Sept	Oct	Nov	Dec	Average for year	No. of years
Mean maximum temperature (°C)	29.4	28.9	28.3	28.3	27.8	27.2	26.1	26.7	28.9	30.6	31.1	30.6	28.3	14
Mean minimum temperature (°C)	18.3	18.3	17.8	17.8	16.1	13.9	12.8	13.9	15.0	16.7	17.8	18.3	16.7	14
Precipitation (mm)	152	109	137	48	5	3	0	0	3	5	23	91	Total for year 577	25

PORT DARWIN, AUSTRALIA: 12°28'S, 130°51'E, 28 m A3 zone

	Jan	Feb	Mar	Apr	May	Jun	Jul	Aug	Sept	Oct	Nov	Dec	Average for year	No. of years
Mean maximum temperature (°C)	32.2	32.2	32.8	33.3	32.8	31.1	30.6	31.7	32.8	33.9	34.4	33.3	32.8	58
Mean minimum temperature (°C)	25.0	25.0	25.0	24.4	22.8	20.6	19.4	21.1	23.3	25.0	25.6	25.6	23.3	58
Precipitation (mm)	386	312	254	97	15	3	0	3	13	51	119	239	Total for year 1492	70

SAN SALVADOR, EL SALVADOR: 13°42'N, 89°13'W, 675 m A3 zone

	Jan	Feb	Mar	Apr	May	Jun	Jul	Aug	Sept	Oct	Nov	Dec	Average for year	No. of years
Mean maximum temperature (°C)	32.2	33.3	34.4	33.9	32.8	30.6	31.7	31.7	30.6	30.6	30.6	31.7	32.2	39
Mean minimum temperature (°C)	15.6	15.6	16.7	18.3	19.4	18.9	18.3	18.9	18.9	18.3	17.2	16.1	17.8	39
Precipitation (mm)	8	5	10	43	196	328	292	297	307	241	41	10	Total for year 1778	39

HOUSTON, USA: 29°46'N, 95°22'W, 12 m BU zone

	Jan	Feb	Mar	Apr	May	Jun	Jul	Aug	Sept	Oct	Nov	Dec	Average for year	No. of years
Mean maximum temperature (°C)	16.7	18.3	22.2	25.6	28.9	32.2	33.3	33.9	31.1	27.2	21.7	17.2	25.6	34
Mean minimum temperature (°C)	6.7	7.8	12.2	15.6	18.9	22.2	23.3	23.3	21.1	16.1	11.1	7.2	15.6	36
Precipitation (mm)	89	76	84	91	119	117	99	98	104	94	89	109	Total for year 1169	45

SYDNEY, AUSTRALIA: 33°52′S, 151°12′E, 42 m BU zone

	Jan	Feb	Mar	Apr	May	Jun	Jul	Aug	Sept	Oct	Nov	Dec	Average for year	No. of years
Mean maximum temperature (°C)	25·6	25·6	24·4	21·7	18·9	16·1	15·6	17·2	19·4	21·7	23·3	25·0	21·1	87
Mean minimum temperature (°C)	18·3	18·3	17·2	14·4	11·1	8·9	7·8	8·9	10·6	13·3	15·6	17·2	13·3	87
Precipitation (mm)	89	102	127	135	127	117	117	76	74	71	74	74	Total for year 1 183	87

PORT ELIZABETH, SOUTH AFRICA: 33°59′S, 25°36′E, 55 m BU zone

	Jan	Feb	Mar	Apr	May	Jun	Jul	Aug	Sept	Oct	Nov	Dec	Average for year	No. of years
Mean maximum temperature (°C)	25·6	25·6	24·4	22·8	21·7	20·0	19·4	20·0	20·0	21·1	22·2	23·9	22·2	14
Mean minimum temperature (°C)	16·1	16·7	15·6	12·8	10·0	7·2	7·2	8·3	10·0	12·2	13·9	15·0	12·2	14
Precipitation (mm)	30	33	48	46	61	46	48	51	58	56	56	43	Total for year 576	84

MONTEVIDEO, URUGUAY: 34°52′S, 56°12′W, 22 m BU zone

	Jan	Feb	Mar	Apr	May	Jun	Jul	Aug	Sept	Oct	Nov	Dec	Average for year	No. of years
Mean maximum temperature (°C)	28·3	27·8	25·6	21·7	17·8	15·0	14·4	15·0	17·2	20·0	23·3	26·1	21·1	56
Mean minimum temperature (°C)	16·7	16·1	15·0	11·7	8·9	6·1	6·1	6·1	7·8	9·4	12·2	15·0	11·1	56
Precipitation (mm)	74	66	99	99	84	81	74	79	76	66	74	79	Total for year 951	56

NEW DELHI, INDIA: 28°35′N, 77°12′E, 215 m BS″ zone

	Jan	Feb	Mar	Apr	May	Jun	Jul	Aug	Sept	Oct	Nov	Dec	Average for year	No. of years
Mean maximum temperature (°C)	21·1	23·9	30·6	36·1	40·6	38·9	35·6	33·9	33·9	33·9	28·9	22·8	31·7	10
Mean minimum temperature (°C)	6·7	9·4	14·4	20·0	26·1	28·3	27·2	26·1	23·9	18·3	11·1	7·8	18·3	10
Precipitation (mm)	23	18	13	8	13	74	180	173	117	10	3	10	Total for year 642	75

ROSWELL, USA: 33°24'N, 104°32'W, 1100 m HBS' zone

	Jan	Feb	Mar	Apr	May	Jun	Jul	Aug	Sept	Oct	Nov	Dec	Average for year	No. of years
Mean maximum temperature (°C)	12.8	15.0	20.0	24.4	28.3	32.2	32.8	32.2	28.9	23.3	17.2	12.2	23.3	37
Mean minimum temperature (°C)	-4.4	-1.2	1.7	5.6	11.1	15.6	17.8	17.2	13.3	6.7	0.0	-4.4	6.7	37
Precipitation (mm)	13	15	18	23	28	46	56	56	51	38	20	15	Total for year 378	53

NICOSIA, CYPRUS: 35°09'N, 33°17'E, 218 m BW' zone

	Jan	Feb	Mar	Apr	May	Jun	Jul	Aug	Sept	Oct	Nov	Dec	Average for year	No. of years
Mean maximum temperature (°C)	15.0	15.6	18.3	23.3	28.3	32.8	36.1	36.1	32.8	27.2	22.2	16.7	25.0	40
Mean minimum temperature (°C)	5.6	5.6	6.7	10.0	15.6	18.3	20.6	20.6	18.3	14.4	10.6	7.2	12.8	40
Precipitation (mm)	74	51	33	20	28	10	2	1	5	23	43	76	Total for year 366	64

CAPE TOWN, SOUTH AFRICA: 33°54'S, 18°32'E, 17 m BW' zone

	Jan	Feb	Mar	Apr	May	Jun	Jul	Aug	Sept	Oct	Nov	Dec	Average for year	No. of years
Mean maximum temperature (°C)	25.6	26.1	25.0	22.2	19.4	18.3	17.2	17.8	18.3	21.1	22.8	24.4	21.7	19
Mean minimum temperature (°C)	15.6	15.6	14.4	11.7	9.4	7.8	7.2	7.8	9.4	11.1	12.8	14.4	11.7	19
Precipitation (mm)	15	8	18	48	79	84	89	66	43	30	18	10	Total for year 508	18

LOS ANGELES, USA: 34°03'N, 118°15'W, 95 m BW' zone

	Jan	Feb	Mar	Apr	May	Jun	Jul	Aug	Sept	Oct	Nov	Dec	Average for year	No. of years
Mean maximum temperature (°C)	18.3	18.9	19.4	21.1	22.2	24.4	27.2	27.8	27.2	24.4	22.8	19.4	22.8	70
Mean minimum temperature (°C)	7.8	8.3	8.9	10.0	11.7	13.3	15.6	15.6	14.4	12.2	10.0	8.3	11.1	70
Precipitation (mm)	79	76	71	25	10	3	1	1	5	15	30	66	Total for year 382	53

ALICE SPRINGS, AUSTRALIA: 23°38'S, 133°35'E, 575 m BF zone

	Jan	Feb	Mar	Apr	May	Jun	Jul	Aug	Sept	Oct	Nov	Dec	Average for year	No. of years
Mean maximum temperature (°C)	36·1	35·0	32·2	27·2	22·8	19·4	19·4	22·8	27·2	31·1	33·9	35·6	28·3	62
Mean minimum temperature (°C)	21·1	20·6	17·2	12·2	7·8	5·0	3·9	6·1	9·4	14·4	17·8	20·0	12·8	62
Precipitation (mm)	43	33	28	10	15	13	8	8	8	18	30	38	Total for year 252	30

LIMA, PERU: 12°05'S, 77°03'W, 120 m BF zone

	Jan	Feb	Mar	Apr	May	Jun	Jul	Aug	Sept	Oct	Nov	Dec	Average for year	No. of years
Mean maximum temperature (°C)	27·8	28·3	28·3	26·7	23·3	20·0	19·4	18·9	20·0	21·7	23·3	25·6	23·9	15
Mean minimum temperature (°C)	18·9	19·4	18·9	17·2	15·6	14·4	13·9	13·3	13·9	14·4	15·6	16·7	16·1	15
Precipitation (mm)	3	1	T†	1	5	5	8	8	7	3	2	1	Total for year 44	15

WADI HALFA, SUDAN: 21°55'N, 31°20'E, 125 m BF zone

	Jan	Feb	Mar	Apr	May	Jun	Jul	Aug	Sept	Oct	Nov	Dec	Average for year	No. of years
Mean maximum temperature (°C)	23·9	26·1	31·1	36·7	40·0	41·1	41·1	40·6	38·3	36·7	30·6	25·6	34·4	39
Mean minimum temperature (°C)	7·8	8·9	12·2	16·7	21·1	23·3	23·3	23·9	22·2	19·4	14·4	9·4	16·7	39
Precipitation (mm)	T†	T	T	T	T	0	T	T	T	T	T	0	Total for year 1	39

†T = trace

BOSTON, USA: 42°22'N, 71°04'W, 40 m CU zone

	Jan	Feb	Mar	Apr	May	Jun	Jul	Aug	Sept	Oct	Nov	Dec	Average for year	No. of years
Mean maximum temperature (°C)	2·2	2·8	6·1	12·2	18·9	23·9	26·7	25·6	21·7	16·7	9·4	4·4	14·4	59
Mean minimum temperature (°C)	−6·7	−6·1	−2·2	3·3	9·4	14·4	17·2	16·7	12·8	7·8	1·7	−3·9	5·6	59
Precipitation (mm)	91	84	97	89	79	81	84	91	81	84	91	86	Total for year 1 038	75

London, England: 51°29′N, 0°00′, 45 m CU zone

	Jan	Feb	Mar	Apr	May	Jun	Jul	Aug	Sept	Oct	Nov	Dec	Average for year	No. of years
Mean maximum temperature (°C)	6·7	7·2	10·6	13·3	17·2	20·6	22·8	22·2	19·4	14·4	9·4	7·2	14·4	30
Mean minimum temperature (°C)	1·7	1·7	2·8	4·4	7·2	10·6	12·8	12·2	10·6	6·7	3·9	2·2	6·1	30
Precipitation (mm)	51	38	36	46	46	41	51	56	46	58	64	51	Total for year 584	30

Andermatt, Switzerland: 46°38′N, 8°35′E, 1430 m HCU zone

	Jan	Feb	Mar	Apr	May	Jun	Jul	Aug	Sept	Oct	Nov	Dec	Average for year	No. of years
Mean maximum temperature (°C)	−4·4	−1·7	2·2	5·6	10·0	13·3	15·6	15·0	12·8	8·3	2·2	−3·3	6·1	40
Mean minimum temperature (°C)	−7·8	−7·2	−4·4	0·6	5·0	8·3	10·0	9·4	6·7	2·2	−1·7	−6·1	1·1	40
Precipitation (mm)	104	99	107	112	112	112	122	130	122	132	104	97	Total for year 1353	70

Hengchow, China: 26°56′N, 112°25′E, 65 m CV′ zone

	Jan	Feb	Mar	Apr	May	Jun	Jul	Aug	Sept	Oct	Nov	Dec	Average for year	No. of years
Mean maximum temperature (°C)	6·1	9·4	16·1	19·4	27·8	30·0	34·4	35·0	31·1	22·8	16·7	12·2	21·7	3
Mean minimum temperature (°C)	1·7	5·0	8·3	13·3	20·0	22·8	25·6	25·0	21·7	15·6	9·4	5·0	14·4	3
Precipitation (mm)	69	132	94	193	168	272	84	145	64	84	119	51	Total for year 1475	4

21796

Oklahoma City, USA: 35°29′N, 97°32′W, 380 m CV zone

	Jan	Feb	Mar	Apr	May	Jun	Jul	Aug	Sept	Oct	Nov	Dec	Average for year	No. of years
Mean maximum temperature (°C)	8·3	10·6	16·7	21·7	25·6	30·6	33·3	33·3	29·4	22·8	15·6	9·4	21·7	56
Mean minimum temperature (°C)	−2·2	−1·1	3·9	9·4	14·4	19·4	21·7	21·1	17·2	11·1	3·9	−1·1	10·0	56
Precipitation (mm)	33	25	56	84	130	89	74	69	76	76	51	41	Total for year 804	40

WARSAW, POLAND: 52°13'N, 21°02'E, 120 m CS zone

	Jan	Feb	Mar	Apr	May	Jun	Jul	Aug	Sept	Oct	Nov	Dec	Average for year	No. of years
Mean maximum temperature (°C)	-1.1	0.0	5.0	12.2	19.4	22.2	23.9	23.8	18.3	12.2	4.4	0.0	11.7	25
Mean minimum temperature (°C)	-6.1	-5.0	-2.2	3.3	8.9	11.7	13.3	12.8	8.9	5.0	0.0	-3.9	3.9	25
Precipitation (mm)	30	28	33	38	48	66	76	76	48	43	36	36	Total for year 559	113

MUKDEN, CHINA: 41°48'W, 123°23'E, 40 m CS" zone

	Jan	Feb	Mar	Apr	May	Jun	Jul	Aug	Sept	Oct	Nov	Dec	Average for year	No. of years
Mean maximum temperature (°C)	-5.6	-2.2	6.1	16.1	23.3	28.9	30.6	29.4	23.9	16.1	5.0	-3.9	13.9	10
Mean minimum temperature (°C)	-18.3	-14.4	-6.1	2.8	10.0	16.1	20.6	19.4	11.1	3.3	-5.6	-15.0	2.2	10
Precipitation (mm)	8	8	18	28	69	84	183	170	63	36	28	15	Total for year 710	10

LINCOLN, USA: 40°49'N, 96°42'W, 360 m CS' zone

	Jan	Feb	Mar	Apr	May	Jun	Jul	Aug	Sept	Oct	Nov	Dec	Average for year	No. of years
Mean maximum temperature (°C)	1.1	2.8	9.4	17.2	22.8	28.3	31.7	30.6	26.1	19.4	10.0	3.3	16.7	61
Mean minimum temperature (°C)	-10.0	-8.3	-2.2	5.0	10.6	16.1	18.9	17.8	13.3	6.1	-1.1	-7.2	5.0	61
Precipitation (mm)	15	23	30	63	102	107	99	89	74	51	30	20	Total for year 703	53

SEATTLE, USA: 47°36'N, 122°20'W, ·38 m CW' zone

	Jan	Feb	Mar	Apr	May	Jun	Jul	Aug	Sept	Oct	Nov	Dec	Average for year	No. of years
Mean maximum temperature (°C)	7.2	8.9	11.1	14.4	17.8	20.6	22.2	22.8	19.4	15.0	10.6	8.3	15.0	57
Mean minimum temperature (°C)	2.2	2.8	3.9	6.1	8.3	11.1	12.2	12.8	11.1	8.3	5.0	3.3	7.2	57
Precipitation (mm)	122	94	79	58	46	36	15	18	43	74	122	142	Total for year 849	70

Kucha, China: 41°40'N, 83°06'E, 960 m CF zone

	Jan	Feb	Mar	Apr	May	Jun	Jul	Aug	Sept	Oct	Nov	Dec	Average for year	No. of years
Mean maximum temperature (°C)	−5·6	3·3	15·6	20·6	26·7	30·0	32·2	30·6	27·2	20·6	9·4	−0·6	17·2	27
Mean minimum temperature (°C)	−19·4	−10·6	−1·1	5·6	9·4	12·8	16·1	14·4	8·9	3·9	−6·1	−13·9	1·7	27
Precipitation (mm)	3	3	5	3	3	33	18	8	5	0	0	5	Total for year 86	18

Goose Bay, Canada: 53°21'N, 60°25'W, 45 m DU zone

	Jan	Feb	Mar	Apr	May	Jun	Jul	Aug	Sept	Oct	Nov	Dec	Average for year	No. of years
Mean maximum temperature (°C)	−13·3	−10·0	−3·9	2·8	9·4	16·1	21·7	19·4	15·0	7·2	0·6	−8·9	4·4	10
Mean minimum temperature (°C)	−22·2	−20·6	−14·4	−7·2	0·0	5·6	11·1	9·4	5·6	−0·6	−7·8	−16·7	−5·0	10
Precipitation (mm)	58	58	61	48	53	61	81	71	58	61	64	64	Total for year 739	10

Fairbanks, USA: 64°51'N, 146°43'W, 134 m DS' zone

	Jan	Feb	Mar	Apr	May	Jun	Jul	Aug	Sept	Oct	Nov	Dec	Average for year	No. of years
Mean maximum temperature (°C)	−18·9	−11·7	−5·0	5·6	15·0	21·7	22·2	18·9	12·2	1·7	−11·1	−17·2	2·8	44
Mean minimum temperature (°C)	−28·9	−23·3	−20·0	−8·3	1·7	7·8	8·9	6·7	0·6	−7·8	−20·6	−26·7	−8·9	44
Precipitation (mm)	23	13	18	8	15	33	48	53	33	20	18	15	Total for year 297	35

Omsk, USSR: 54°58'N, 73°20'E, 85 m DS' zone

	Jan	Feb	Mar	Apr	May	Jun	Jul	Aug	Sept	Oct	Nov	Dec	Average for year	No. of years
Mean maximum temperature (°C)	−18·3	−14·4	−7·2	−3·9	15·0	20·6	23·3	21·1	15·6	4·4	−7·8	−15·0	3·3	22
Mean minimum temperature (°C)	−25·6	−22·8	−17·8	−6·1	4·4	10·6	13·3	11·1	5·0	−2·8	−13·3	−21·7	−5·6	19
Precipitation (mm)	15	8	8	13	30	51	51	51	28	25	18	20	Total for year 318	22

STOCKHOLM, SWEDEN: 59°21'N, 18°04'E, 45 m DS zone

	Jan	Feb	Mar	Apr	May	Jun	Jul	Aug	Sept	Oct	Nov	Dec	Average for year	No. of years
Mean maximum temperature (°C)	-0·6	-0·6	2·8	7·2	13·9	18·3	21·1	18·9	14·4	8·9	3·3	0·6	8·9	30
Mean minimum temperature (°C)	-5·0	-5·6	-3·3	0·0	5·0	9·4	12·8	11·7	7·8	3·9	-0·6	-3·3	2·8	30
Precipitation (mm)	38	28	28	38	41	48	71	79	53	53	48	48	Total for year 573	30

REYKJAVIK, ICELAND: 64°09'N, 21°56'W, 30 m DW zone

	Jan	Feb	Mar	Apr	May	Jun	Jul	Aug	Sept	Oct	Nov	Dec	Average per year	No. of years
Mean maximum temperature (°C)	2·2	2·8	3·9	6·1	10·0	12·8	14·4	13·9	10·6	6·7	3·9	3·3	7·8	30
Mean minimum temperature (°C)	-2·2	-2·2	-1·1	0·6	3·9	6·7	8·9	8·3	5·6	2·2	0·0	-1·1	2·8	25
Precipitation (mm)	102	79	76	53	41	43	51	66	79	86	91	94	Total for year 861	30

POINT BARROW, USA: 71°18'N, 156°47'W, 7 m E zone

	Jan	Feb	Mar	Apr	May	Jun	Jul	Aug	Sept	Oct	Nov	Dec	Average for year	No. of years
Mean maximum temperature (°C)	-22·2	-24·4	-22·2	-13·9	-4·4	3·9	7·8	6·7	1·1	-5·6	-13·9	-20·0	-8·9	32
Mean minimum temperature (°C)	-30·0	-31·7	-30·0	-22·2	-10·6	-1·7	0·6	0·6	-2·8	-11·1	-20·6	-27·2	-15·6	32
Precipitation (mm)	5	3	3	3	3	8	23	20	13	13	8	5	Total for year 104	32

NOVAYA ZEMLYA, USSR: 72°23'N, 52°43'E, 15 m E zone

	Jan	Feb	Mar	Apr	May	Jun	Jul	Aug	Sept	Oct	Nov	Dec	Average for year	No. of years
Mean maximum temperature (°C)	-16·7	-16·1	-13·9	-8·9	-2·8	2·2	7·2	7·2	2·8	-2·8	-10·6	-14·4	-5·6	29
Mean minimum temperature (°C)	-20·6	-20·0	-19·4	-14·4	-7·2	-1·1	3·3	4·4	0·0	-5·6	-13·9	-17·2	-9·4	17
Precipitation (mm)	8	8	8	8	15	20	36	43	43	33	13	12	Total for year 247	18

Chapter 7

MICROCLIMATES

7.1 Introduction

In the previous chapters only the broad patterns of the macroclimate of the globe, measured in relation to standard conditions of exposure over an area that seeks to reduce the effects of the immediate environment to a minimum, have been discussed. However, for most applied problems knowledge of the macroclimate is quite insufficient and the investigator has to turn to a study of the microclimate of the particular habitat in which he is working. It surely has been made clear to everyone what a difference local conditions can make to the climate: on a hot day, just walk across the tarmac surface of a parking area and then compare it with the conditions experienced standing on grass under a nearby tree.[1] Clearly the reflection, the absorption, heat capacity and other physical properties of the environment all play their part in determining the microclimate of the area.

The most comprehensive publication dealing with measurements made in various microclimates is *The Climate Near the Ground* by Geiger.[2] It is essential reading for anyone working in this field. Within the next few pages we will discuss some of the more important aspects of microclimates variation of selected climatic parameters and trust that with the understanding gained and the application of common sense, supplemented by measurements, an investigator will be able to appreciate some of the difficulties with which he is dealing.

7.2 Radiation, Sunshine, Light and Temperature

Radiation energy is, as has been pointed out, greatly affected by the angle at which it strikes a surface. In the northern hemisphere, a north-facing slope will receive much less radiation than a south-facing one, until, for example a north-facing slope in the approximate latitude of London, at an angle of more than 40° inclination, will receive no direct sunlight between the autumn and the spring equinoxes. Even during the summer direct radiation will only fall on such a slope between 9.0 a.m. and 3.0 p.m. (Fig. 7.1). This knowledge has been utilized very successfully in the Rhine valley where vineyards are cultivated on steep slopes facing towards the south and south-west in latitudes were the radiation received on a horizontal surface would not be adequate for the crop.

In addition to the degree of slope and aspect, the amount of shading of surrounding objects is fundamental. For example, in narrow city streets the radiative pattern is very different in an E-W road from a N-S one.[3] Thus, the local environment must be known in some detail if estimates of the radiation are to be made or if the records obtained from a local station are to be modified by sensible techniques.

Of course, in some areas, such as within crops, woodlands and the like, the quality, or spectral distribution, of the radiation is also altered by its absorption and reflection by vegetative matter.

Sunshine and light are dependent upon the same basic factors as the radiation and the previous comments will also refer to these elements. It must be pointed out though that some organisms are responsive to wave-lengths of energy different from those that influence human vision; the acceptance of readings of light may therefore lead to incorrect deductions.

Temperature is closely related to the radiative energy. It differs from sunshine and light in that with the advection of a different air mass into a

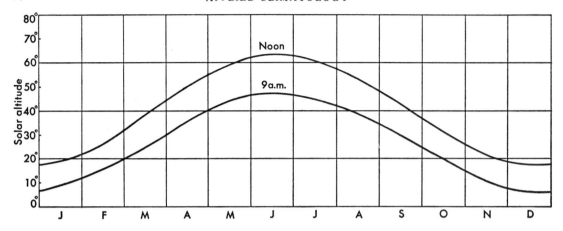

Fig. 7.1 Solar altitude in lat. 50 N at noon and 9.0 a.m. and 3.0 p.m.

region little change in radiative load may occur while a change in temperature is brought about. Special note should be made of such phenomena as land and sea breezes, katabatic flow and foehn winds (3.7). These may affect the microclimate of the site but not that of the observing station a few metres or kilometres away, and vice versa.

7.3 Humidity and Evaporation

The absolute amount of water vapour in the air is naturally influenced by the quantity of water available. Around vegetation and water surfaces, unless the former are in a wilting condition, the humidity will be greater than in the open. However, the amount of transpiration from plants is dependent upon the behaviour of the stomata and these are generally more active during the daylight periods so that the source of transpired moisture will reduce or cease at night. The surface of the ground, assuming that there is water within the top few centimetres of the soil, will provide another source of water vapour, and this moist air may be soon lifted into the higher levels by convective forces.

The evaporation of water from whatever source, has the effect of cooling the air, so it is to be expected that within fairly dense vegetation or around water surfaces the air will be cooler but with a higher relative humidity. Over relatively open ground, in spite of the fact that the evaporation of soil moisture may use some of the radiant energy, there is still usually sufficient left for high temperatures to be reached.

It must be realized that although a large amount of moisture may be evaporated, the increase in the overall humidity is likely to be small unless the air flow is low. When calm conditions prevail the high humidity and radiation cooling at night often lead to the formation of mists over fields and other moist areas. Wind speeds of, say, 15 km/hour will soon cause a great degree of mixing with the upper air to take place and the increase in humidity will become negligible. If, under a high wind flow, there is to be an effect it must usually come from a long traverse of the wind over an open water surface.

7.4 Air Flow and Precipitation

It has been indicated how the air speed varies with height and how the direction may reverse completely because of the pattern of the vegetation (2.2.8). This means that if wind velocity is an important parameter in an investigation there is very little to do except to measure it at the places under investigation.

The air flow will play a part in many branches of applied climatology for, as just noted, it will affect the humidity stratification, the temperature, especially that of any surface where evaporative cooling takes place, and the precipitation. Geiger[2] has noted that on a hill, a protrusion not high enough to cause thermodynamic considerations to be applied, the slope on the windward side receives less precipitation than the leeward side. On the windward side the precipitation is carried away by the wind until it reaches the region of gentler winds and can then fall out. Of course under conditions of intense rain and wind speed the obliquely-falling drops are often beaten downwards and the difference is partly nullified.

Rainfall within a forest may present a very different pattern from that in the near-by open exposures because, in a dense stand, the rain may take a long time to penetrate through the canopy. It will then manifest itself in the form of large droplets falling from leaves or rivulets of water coursing down the trunks and stems of the vegetation. The only way to get a good estimate of the rain falling on to a woodland is to erect a rain gauge just above the canopy leading the collected water down to a container at ground level. However, although this may yield a figure useful for catchment calculations it will not necessarily give a good idea of the conditions within the wood.

7.5 Modifications of Climate or Weather

The four main methods of modifying the climate of a relatively small area that will be discussed here are windbreaks, artificial stimulation of rain, the reduction of evaporation and the preventing of frosts. The remarks here are not intended to signify that all of these methods are proved successful although, with certain reservations, it is likely that such modifications can be made if conditions are just right.

7.5.1 Windbreaks

Historically the construction of windbreaks probably constitutes the earliest attempts to modify weather. They have proved effective in controlling drifting snow and the reduction of wind pressure on objects appreciable distances from the nearest break or shelter.

It has been shown that if the height of the tree belt is H, then for a distance of about $40H$ there is a significant reduction in the air speed, while within about $5H$ there is almost calm, with some small independent cell circulation (Fig. 7.2). Of course, for a shelterbelt to be reasonably effective it must be at right angles to the prevailing wind direction, and this presupposes that there is a predominant direction of air movement from which the site must be protected.

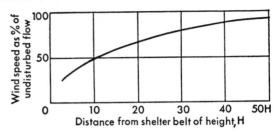

Fig. 7.2 Effect of a shelterbelt on wind speed (after Jensen)

A shelterbelt reduces air flow, causes increased shading of the near-by crops, changes the precipitation pattern, especially of rain coming from the prevailing direction, and also changes the evaporation taking place from crops and soil. It is not possible to say whether in general the belt will cause an increase or decrease in the evaporation amounts.[4]

Shelterbelts can of course be used to protect certain areas such as farms or fields from blowing and piling snow, but care must be taken in construction to make sure that all directions from which snow may blow are anticipated, otherwise the presence of trees may increase piling in unexpected places. It is interesting to note that, practically, the best results of sheltering seem to occur when the tree break presents about a 40 per cent coverage in the vertical plan. The reader specifically interested in the many aspects of this problem is referred to a very comprehensive publication by Jensen.[5]

Trees may be used to shade delicate crops from the intense sun's rays during certain stages of their growth. Such shading was standard practice in tea and coffee plantations, but experiments suggest that it may not be necessary.[6]

7.5.2 Artificial stimulation of rain

It has been demonstrated that by introducing artificial nuclei into suitable clouds a release of precipitation may be induced. Mason[7] has stated that the deficiency of suitable agents in some clouds can be remedied by seeding them with solid carbon dioxide, silver iodide or water droplets or other large hygroscopic nuclei. The solid carbon dioxide or silver iodide produce ice crystals and thus act as freezing nuclei; the water droplets or hygroscopic nuclei such as dry, finely-powdered salt, act as condensation nuclei Hygroscopic nuclei are advised for 'warm' clouds, i.e., those that are mostly at a temperature above freezing point.

Various methods have been tried for getting the nuclei into the clouds. These include the use of burners so that the ascending hot air carries the particles upwards, rockets that are timed to explode at or near the base of the cloud and direct seeding from aircraft. The base of the cloud is chosen because it is here that vertical currents are usually sufficiently well developed for the particles to become satisfactorily circulated within the cloud cell.

Not all of these methods have been developed fully, but at present it seems advantageous to use aircraft so that the investigator may be certain the particles are being released in the correct place.

Reports of results vary, mainly because it is extremely difficult to judge whether the seeding operation has caused the precipitation or whether it would have rained in any case. Because of this many tests have to be carried out and the results subjected to statistical analysis. Often slight changes in the method of analysis can lead to differing results, viewed from the point of mathematical significance. Some experiments purport to show an increase in rainfall, some no significant change and others a decrease. However, the increase noted is, at best, of the order of 10 per cent. The author has taken part in many trials in East Africa and, although initially a sceptic, is now of the opinion that if clouds are on the point of beginning precipitation that would reach the ground the addition of the nuclei can speed up the process by the order of 10 minutes. Such a speeding up can be very important when the result is rain on farmed plains instead of on un-farmed mountains (Fig. 7.3).

7.5.3 *Reduction of evaporation*

Evaporation results in an enormous loss of water to the atmosphere. For example, 30 cm of evaporation from a 1 hectare pond per year is equivalent to the loss of about 4 million litres. Thus, in areas where water is in short supply it is very desirable to find some method of reducing evaporative loss.

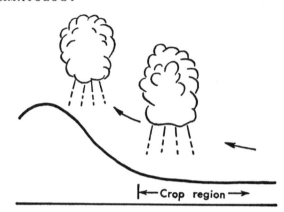

Fig. 7.3 How seeded cloud may precipitate a little earlier and on an economically more important area

The idea of floating a monomolecular layer of some completely harmless chemical on the surface, to act as a cover, was first mooted about 40 years ago. In the past 20—25 years many experiments using cetyl alcohol, a mixture of duodecanol and hexadecanol hydrocarbons, have been made. The chemical is quite harmless but has a low surface tension, breaking at a pressure of some 24 dyne/cm. This physical characteristic appears to the author to be the real crux of the matter for, in many experiments in East Africa, it was found that whenever a wind speed, measured about 25 cm above the water surface, in excess of some 5 km/hour was noted, the film broke. For it to remain unbroken, the solution, either as an emulsion or in kerosene, had to be dispensed continuously. Changes in wind direction also gave problems because the position of the dispensers had to be changed unless they were sited all round the perimeter of the pond. Because of the cost of the chemical it was economically desirable not to over-dispense the solution. In addition, excess lateral surface pressure caused the powder to deposit out, necessitating almost continuous supervision of the dispensing process by adequately trained personnel.

Experiments in the laboratory and with small evaporation pans in the field gave results of up to 70 per cent saving in evaporative loss, but well-documented field experiments on large bodies of water set the practical saving at about one-tenth of this. It is the opinion of the author that the method can prove practical on small surfaces of water, up to about 0·1 hectare, where, by the erection of windbreaks of some form, the air flow can be kept low. But on a large surface it would be necessary to use some chemical with a much greater surface tension.

A book edited by La Mer[8] gives an excellent review of the physics and experiments involved in this problem.

7.5.4 *Frost prevention*

Frosts are of two types, radiation and advection. The conditions in which the former are most likely to form are calm, dry, clear air, because the ground and air can then lose heat by radiation to the cold night sky. In the latter type the temperature is lowered because of the movement into the area of a different, colder, air mass. Thus, the latter phenomenon generally takes place on almost a macroclimatic scale and man's efforts to counteract it have so far proved unsuccessful.

In the case of a radiation frost, however, the following precautionary methods can be used.

(1) Because a cloudless sky gives rise to the greatest heat loss an artificial cloud can be formed so as to blanket the area. This method is used when the smudge pots, burners of cheap smoky kerosene, are lighted since these cover the area with a grey-black smoke.

(2) As the radiative cooling is most pronounced near to the ground there exists a level, usually about 30 m, in the atmosphere where the temperature is warmer. Owing to the cold air being more dense the actual conditions are stable but if artificial mixing can be introduced some of the warmer air can be brought downwards.

This technique is used when giant fans are switched on since these mix the thermal stratification so helping to warm the lower layers.

(3) Since the vegetation and air are losing heat through the radiation process, if heat can be introduced from some other source the actual temperature drop can be lessened. One method at present tried for putting heat into the cycle is to sprinkle the vegetation with fine water droplets; these condense and latent heat is released; also, when the droplets freeze more heat is released. However, the amount of water and its application are critical: if too much is applied the ice formed can become too thick, freezing the vegetation or destroying it by sheer weight. Another method is via the smudge pots which do create heat within the cold volume. Unfortunately, the heat is rapidly dispersed due to the natural air flow and some form of windbreak may increase the efficiency of the method.

Some types of burners that give off little or no smoke, relying on direct heating and resultant convection to increase the temperature are in common use.

(4) An efficient method is to cover plants with caps or cloches of plastic, treated paper or similar materials. On calm nights they act as radiation shields and can keep temperatures beneath them some 1·6–3·2 °C above the unshielded temperature at the same level. During the day the caps cause a rise in temperatures beneath them. Due to the cold night-time and warm daytime temperatures of the caps the plants should not be in contact with them. A method using foam to cover short crops has proved effective in trials.

Chapter 8

CLIMATE AND SOILS

8.1 Introduction

Soil, earth or ground is generally defined as a mixture of disintegrated rocks and organic material in which plants are rooted. The fact that it is not the original ground is embodied in the definition which adds: 'it is the matter of the surface of the globe.' Soil is simply a detritus, but a most important one to mankind.

Soil is the end-product of a number of factors all working together:

(*a*) parent material or bedrock;
(*b*) climate;
(*c*) relief;
(*d*) plant and animal life; and
(*e*) time.

Of these five, three are either completely or partly functions of the climate, for (*c*) ultimately affects the microclimate of the region and (*d*) reflects the effect of climate indirectly, acting through the type of plant or animal existing in that region.

Soil is considered to show various layers, or horizons. Beginning at the surface these are:

topsoil: in which organic debris, leaching and eluviation are maximal;
subsoil: maximum accumulation of clays, deeper-coloured;
weathered bedrock;
underlying bedrock.

These layers are sometimes referred to as horizons A, B, C, D respectively. Not all horizons are easily identifiable or present in every soil.

Leaching (the removal of mineral or organic compounds in solution) and eluviation (the transport of small colloidal solids) are plainly influenced by the climatic elements of which rainfall, its amount and intensity, is the most important. While we are considering definitions, clay consists of particles less than 0·005 mm diameter; silt or those with diameters between 0·005 mm and 0·05 mm; and sand and gravel particles up to 2·0 mm in diameter. Using these three variables many different soil textures may be derived, a major one being loam (50 per cent sand, 50 per cent silt).

Humus, the decayed organic matter in the soil, is a colloidal brown substance that assists in the formation of solutions so that plants can utilize certain substances. Its formation and rate of formation are greatly affected by the climate; for instance, if the soil becomes waterlogged, air within the soil interstices is greatly reduced, the number of bacteria decreases and the plant refuse remains as simple acid material.

The alkalinity or acidity of soil is measured in terms of its pH value (a measure of the amount of hydrogen in the soil). Neutral soil gives a pH of 7·0, acidic soils range, normally, from 4·5 to 6·9, and alkaline soils have values above 7·0. As a general rule crops grow best at a pH of 6·0 to 6·5, for, if the soil is too acidic, the essential bacterial action is reduced while alkalinity prevents the plants from using certain trace elements in the soil.

Freezing temperatures also play an important part in the soil structure in certain areas. For instance, if a soil to which water is being continually fed from the groundwater reservoir is subjected to freezing temperatures, ice layers grow continually and the ground surface rises. This frost heave, as it is called, can be considerable (up to 15 cm or more) and, upon thawing, the area affected is transformed into a marshy region. In some other areas a regional freeze-thaw cycle occurs, for instance in tropical mountains above about 4 000 m there is a 24-hour freeze-thaw cycle. This process apparently sorts the particles into relatively uniform size groups.

Soil types are generally divided under three main headings.[1]

Zonal: easily-identifiable horizons resulting from appreciable climatic and biological influence; a definite correlation with climate.

Intrazonal: the bedrock and relief dominate. Note that microclimatic effects may play their part here, e.g. marshes due to relief.

Azonal: horizons not well differentiated, the time factor being too short, e.g. alluvium or sand dunes.

8.2 Zonal Soils

The bedrock has surprisingly little influence upon the zonal soil for it is the processes, such as leaching or eluviation, which are the main determinants. Such processes are related to the climate so that a granite bedrock in the tropics will yield a soil quite unlike that from a granite bedrock in a cold climate.

Before the types of soil related to climatic zones are discussed, it is essential to define certain terms that must be used here:

laterization: the rapid removal of silica under the action of high temperature and abundant precipitation;

podzolization: the removal of iron and silica from the top (A) horizon;

podzol: a highly acidic soil with a peaty surface layer;

podzolic soils: not quite as acidic, relatively little organic matter at the surface;

prairie soils: much organic matters in A horizon, less leaching and eluviation with less rainfall, fertile;

chernozems: much organic matter in A, lime accumulation lower, very fertile;

chestnut and brown soils: from steppes, less organic matter, lime accumulation nearer the surface, slightly alkaline;

sierozems: indefinite horizons, little humus, lime near the surface.

Using these definitions and, by interpreting the results from a climatic viewpoint, it is possible to obtain the following breakdown.

Hot Zone

(*a*) rain forest and wet savanna: high degree of laterization, alkalis eluviated causing acidic soil, humus content low, not very fertile, usually red in colour;

(*b*) tropical grassland: much humus, more fertile but rapidly exhausted, dark colour;

(*c*) deserts: little organic matter, lime accumulation near surface.

Warm Zone

(*a*) Mediterranean regions: leaching reduced, lime accumulation deeper;

(*b*) eastern regions: usually considerable laterization, scanty organic material;

(*c*) deserts: as Hot Zone deserts.

Cool and Cold Zones

(*a*) wetter areas: podzolic soils, thin layer of humus;

(*b*) medium rainfall (prairies): high humus content slight eluviation, fertile;

(*c*) less rainfall (steppes): deep humus layer, lime accumulation deep down, retain water, very fertile;

(*d*) short summers (tundra): anaerobic, little humus, very acid.

It is clear from this short consideration of the major soil types that climate plays an important role in the development of soil, and Fig. 8.1 shows how this may be depicted schematically.

8.3 Intrazonal and Azonal Soils

Intrazonal soils show little dependence upon climate although some relationships can be noted. For example, the saline and alkaline soils (halomorphic) often occur in arid regions where intense evaporation soon removes any surface water. These soils are unsuitable for crop raising until the salts are washed away. The hydromorphic soils, characteristic of poorly-drained regions such as swamps and bogs, can be found in regions where the relief is such that rainfall runoff and seepage is concentrated into a relatively small area.

Azonal soils, which have not had sufficient time to develop definite horizons, are rarely related to climatic conditions, although the regosols, such as loess, occur in certain regions because they were transported by wind and then washed out of the air by rain. Also, alluvial soils are found along river beds, their extent and depth generally being dependent upon such river conditions as the amount and rate of flow, which are themselves functions of the hydrologic cycle.

8.4 Climate, Soil and the Standard Continent

Owing to the dependence of soil upon climate, mainly temperature through the frost free period, and rainfall, via precipitation effectiveness, it is possible to construct a diagram showing soil v. climate (Fig. 8.1). Having set up the relationship between climate and soils, it is now a simple matter to transfer this pattern to the climate obtaining on the standard continent. Also shown, in parentheses, in Fig. 8.1 are the new names of the soil orders — a system that attempts to classify soils rather than the soil forming processes.[2]

In conclusion it should be mentioned that this chapter has been concerned with climate and soil formation, but climate can also play its part in soil destruction, for instance, through soil erosion.

Soil erosion is a curse to which much of the world's surface is prone. Weather factors play a large part in its magnitude and importance, especially when the natural cover of vegetation has been removed. There are three main types of erosion occasioned by climatic factors, these are:

(*a*) wind erosion, wherein the exposed usually

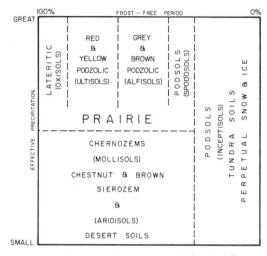

Fig. 8.1 Temperature, rainfall and soil

fertile topsoil is blown away by intense air movement;

(*b*) 'sheet' erosion, wherein heavy-intensity rainfall causes a washing away of soil under the effect of gravity;

(*c*) 'gully' erosion, in which local relief, or occasionally cultivation methods, leads to a concentration of water after a storm, a concentration that washes out soil and rocks and may destroy adjacent land.

In a paper by Wischmeier and Smith[3] the authors suggest a soil loss equation that contains a rainfall factor determined by the storm's kinetic energy and its maximum intensity in 30 min. In the eastern U.S.A. (east of 104 °W) this value ranges from 50 units in western North Dakota to over 600 units in southern Louisiana. An FAO publication[4] gives interesting observations on the relationship of wind to soil erosion.

Chapter 9

CLIMATE AND VEGETATION

9.1 Introduction

It is not the purpose of this chapter to discuss evolution in the plant world or to try to enumerate, let alone evaluate, the various factors that go to make up the development of a certain plant complex or community. However, it is necessary to indicate just how important a part is played by climate in the determination of the type of vegetation that grows in a specific area.[1] The previous chapter has indicated how the type of soil is, to a large degree, a function of the climate and the vegetation; this chapter will try to show the relationship between vegetation, climate and soil. It is not possible to decide just how interrelated are the parts played by soil and vegetation, but the common factor always comes out to be climate.

It is again necessary to define a few terms in general use.

Hygrophytes: plants living in water or very moist climates;
xerophytes: plants adapted to withstand droughts;
tropophytes: plants able to adapt themselves to either condition;
mesophytes: plants needing an average amount of water;
epiphytes: plants that draw moisture from the air and thus need high relative humidity.

The xerophytic plants generally use one of three methods in order to overcome the long periods of drought:

(*a*) the vegetation has thick, waxy bark and leaves which inhibit great transpiration loss;
(*b*) like cacti, they store water within the plant;

(*c*) the plant's roots go down very deep, with a pronounced spreading to assist in the location of underground moisture.

Most plants cease their growth when the soil temperature is below about 6 °C, for, if the temperature is too cold, there is a low rate of moisture intake and the plant cannot replace the transpiration loss quickly enough. Freezing temperatures can damage the plant cells by causing desiccation and chemical changes. High temperatures cause increased evapotranspiration until the plant wilts, but rarely, except by 'burning', cause trouble directly themselves.

9.2 Forests and Woodlands

As with all scientific descriptions, it is essential to make certain that definitions are standardized, especially as the terms used in describing vegetation communities[2] have entered everyday language and become modified or transformed.

Forest: an uncultivated tract of land covered with trees in close proximity, often with more than one canopy.

Woodland: land mainly covered with trees, crowns of which more less touch.

Jungle: strictly means wasteland but often used to describe a dense mass of vegetation; a tropical forest with much undergrowth.

Naturally, areas of forest and woodland blend into each other and it is often difficult to set down accurate word pictures describing the various types. Broadly, it is possible to recognize nine different types of forest and woodland.

(1) *Equatorial evergreen forest (rain forest).* Although called evergreen, the leaves are shed,

albeit sporadically, such that they preserve a pattern of continuous greenery. Typical trees include the ebony, mahogany, rubber and sapele, and lianas (woody vines) and epiphytes abound. Where man has influenced the balance or, on the margins, where sunlight can penetrate, dense undergrowth abounds and a 'jungle' results. In clearings the giant monocotyledons, such as banana, plantain and ginger, thrive. With increased elevation conifers, cedar and juniper, appear.

(2) *Tropical semi-deciduous forest.* Sometimes called the monsoon forest. The region has a seasonal rhythm and includes some leaf shedders, such as teak. In Africa it corresponds to the miombo woodlands (*Isoberlinia* and *Brachystegia*).

(3) *Tropical thorn woodland* (often wrongly called 'forest'). Usually found in regions with a pronounced drought period and includes the acacia (*Mimosa* family) in its many varieties.

(4) *Sclerophyllous woodland.* The adjective means, simply, that the vegetation has a hard, stiff leaf which combats drought by reducing transpiration. Such vegetation is generally found in the 'Mediterranean' climates where there is a long summer drought. The woodland usually includes a number of conifers, some evergreen dicotyledons (such as the live oak), palms, shrubs (such as azalea, rhododendron, laurel) and eucalyptus.

(5) *Mesophytic woodland.* Typical of the subtropical regions with some rain the year round. Palms, deciduous trees, some conifers, acacia, tree ferns, magnolia and camellia grow here.

(6) *Tropophytic woodland.* A temperate deciduous woodland including conifers with some deciduous trees (elm, beech, ash, oak, sweet chestnut). It shows luxuriant summer foliage and has a more rapid metabolism.

(7) *Northern forests.* These comprise mainly conifers (pine, fir, spruce, larch, cedar) with a small percentage of deciduous trees (beech, aspen, willow, poplar). These are characterized by a marked stunting of leaves and/or trunk, especially at the poleward limits.

(8) *Montane forest.* This mainly occurs in tropical and subtropical highland regions with year-round rainfall; sometimes called cloud forest. It usually contains enormous numbers of epiphytes, mosses, lianas, tree ferns and includes bamboo 'forests'.

(9) *Water forests.* The mangrove swamps of the tropics and the bald cypress swamps of the subtropics are the only two examples.

9.3 Shrubland and Grassland

A shrub, or bush, is simply a low, woody plant with little or no trunk. Grass is any plant of the monocotyledonous family Gramineae, which includes wheat and other cereals, bamboo and sugar cane, among others. We use the term grassland to connote an area in which pasture grasses are available at some time of the year for a prolonged period, at least of the order of a few months. Basically there are seven types of shrubland and grassland.

(1) *Tropical grassland.* The savanna, llanos or campos. A long drought period occurs in these regions but the intense rain that occurs over a short period gives rise to very tall grass (elephant grass), and trees such as acacia and baobab.

(2) *Temperate grassland.* The prairies, with medium-length grass cover.

(3) *Temperate grazing grassland* (meadow type). This type differs from the prairies in that it grows in a region where there is fairly uniform rainfall. The grass is of short to medium length and, where the annual amount is about 500–750 mm, produces some of the best grazing land in the world.

(4) *Cold temperate grassland.* Where the rainfall is less, the steppe areas of short grass (buffalo and grama grass) are found. There is summer rainfall, but only in small amounts.

(5) *Cool temperate grasslands and heath.* Barren open country, usually with poor soil and many ericaceous shrubs, such as heather, ling and bilberry, and ferns. In some cases poor drainage makes for marshy conditions. When well drained, yet moist, with fertile and warm soil, these areas tend to yield lush alpine or mountain meadows.

(6) *Montane shrubland.* The ericaceous belt; in the tropics giant lobelias and groundsel thrive.

(7) *Scrubland.* A dense mass of low-growing evergreen plants with occasional trees. It is a region of poor rainfall, or poor soil, and is typical of the semi-arid regions on the fringes of the hot deserts. Examples are the chaparral, mallee scrub and the maquis. Similar to the bushland and

thicket of tropical areas, it is sometimes so dense that it is impenetrable to man.

9.4 Deserts

Although, with the usual climatic definitions, deserts cover a large percentage of the land surface (ca. 20 per cent), only a small part of this is completely barren. Desert areas may be considered to have four subdivisions.

(1) *Desert grass and shrubs.* A region of xerophytic shrubs with some grass and large bare patches.

(2) *Desert scrub.* Dry, stunted woodland, consisting of woody plants, broad-leafed and deciduous, with much bare ground.

(3) *Desert.* Barren, with no vegetation.

(4) *Tundra.* Mosses and lichens, some flowering plants, land often swampy; similar to the paramos, the tropical highland areas above the tree-line and below the snow-line.

9.5 Climate, Vegetation, and the Standard Continent

It is obvious that climate plays a decisive role in deciding the flora of an area. When the vegetation complex has become stabilized it is often referred to as the climatic climax vegetation. These 'end

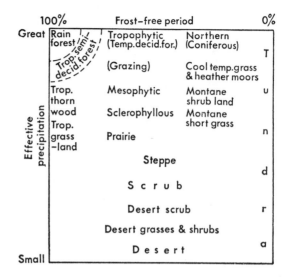

Fig. 9.1 Temperature, rainfall and vegetation

results' are called biomes. The transitional zones between the biomes are named ecotones.

Because of the vegetation dependence on two main elements, precipitation and temperature, these two factors can be utilized to give a schematic diagram of the climate-vegetation relationship (Fig. 9.1). This relationship, whilst not being perfect, does enable a simplified vegetation distribution for the standard continent to be derived.

Chapter 10

CLIMATE AND AGRICULTURE

10.1 Introduction

In the same way that climate helps to define the climax vegetation formation it sets limits for crop production. Of course, crops are also influenced by soil, relief, insects, diseases, etc., but no crop can achieve real importance in an agricultural system unless it is well adapted to the existing environmental conditions, allowing for any necessary irrigation.

The main climatic elements in agriculture are temperature, moisture — in all its forms, sunlight, winds and evaporation. The crop that is being grown must find suitable patterns of these within its microclimate or fail as an economic asset. Man can, by certain practices, control or alter the climate of small localities to make the raising of a certain crop a possibility, but the methods are usually costly. There are some exceptions to this, such as irrigation, windbreaks, smudge pots, mulches, etc., but other methods, such as the covering of fields with plastic sheets to reduce evaporation loss, are very expensive. All crops have certain natural threshold limits of the climatic elements beyond which they do not grow normally but breeding and selection are gradually extending these thresholds for many crops, such as wheat, barley and corn. The methods used during cultivation also play their part, sometimes through the introduction of a favourable change in the microclimate.

The biggest problem in agricultural meteorology today is the effect of the integrated climatic elements upon the crop. As Parker[1] expresses it:

Plant growth is dependent upon all the factors that make up the environment. No particular level of one factor should be referred to as the optimum for growth of a species without specifying at least the approximate levels or conditions of the other important components of the environment. No single optimum for any factor exists that will hold without regard to the status of the other factors. In other words, no single factor in the environment produces its effects simply, but this effect involves interaction of a high order with those produced by other factors. It will be necessary, therefore, to study the interrelationships, the interactions, of all possible factors in the environment before it will be possible to define the optimum 'climate' for the growth of a specific plant or organism.

Many climatic chambers have been built to test the effect upon crop growth and yield of varying one or more elements, but we have still a long way to go before we know the optimum conditions for the production of a certain crop. However, important additional information can be obtained by the research worker if it is understood that, in most cases, it is the immediate environment of the crop, its microclimate, that is fundamental, not the climate of the meteorological shelter. The investigator must make comparisons between the 'standard' climate and the crop climate if ever he is to understand the interrelationships between climate and agriculture.

The practical problems of agrometeorology are further complicated by the two important aspects of plant diseases and pests. Many of the diseases and pests flourish when certain, usually unknown, climatic thresholds are reached. Other diseases and pests await the weakening of the plant to a stage at which it is particularly vulnerable to attack. This weakening is often brought about by climatic conditions. Even within the soil, climatic conditions are of prime importance because of the part they play in the activities of animals such as nematodes (worms). A review of some of the principles and problems of using meteorological data to forecast the trend of diseases and pests is given in an article by Bourke.[2] The problems of crop adaptation and distribution have been discussed by Willsie.[3] In his book he treats the various environmental factors and considers the distribution of important crops on a climatic basis.

The journal *Agricultural Meteorology* is an important source of information for workers in this field.

10.2 Temperature

All crops have minimal, optimum and maximal temperature limits for each of their stages of growth. These limits can vary appreciably; for example, tropical crops, such as cacao, dates, etc., need a high temperature throughout the year while winter rye can withstand freezing temperatures during its long winter dormant period. In general, a high temperature is not so destructive as a low temperature, provided that the moisture supply is sufficient to prevent wilting. It must be realized, however, that the optimum temperature for the maximum growth rate is not necessarily the same as that for optimum crop production.

Some plants are very susceptible to high temperatures when young, although they later can stand extreme heat. Plants of this type, such as tea and coffee, are artificially shaded by other, mature, trees or cloth at this stage. In some cases they are also planted on slopes known to be less exposed than others. It is now suggested that these crops do not need this shade and can grow more productively in open exposures.[4]

High temperatures can sometimes lead to sun-scald, an injury caused by high radiation which leads to excessive heating of plants openly exposed to sunlight. It can also occur during winter, when parts of a plant experience large temperature variations on the side exposed to the sun in cold weather; for this reason tree trunks are sometimes whitewashed. Sun-scald is also the name given to the damage produced on orchard fruits during periods of frosty nights and sunny days.

Large bodies of water, such as the great lakes of America, the Black Sea and the Caspian Sea, are instrumental in alleviating the extremes of temperature in narrow bands, a few tens of miles wide, along the leeward sides. This moderation in the temperature can help, for instance, to prevent blooming until the danger of a killing frost is past.

Certain plants can be killed by chilling temperatures. These are temperatures that are low, but above freezing level. They cause a reduction in the water flow to the roots of the plant, which then wilts and dries up. Two or three such days can kill rice and cotton, while potatoes, maize and many vegetables can withstand such a period. Care must be taken when applying cold irrigation water because its contact with the plant can produce a chilling effect. This can be exaggerated by an additional lowering of the temperature caused by the evaporation of the surface water. If the soil is kept wet, however, it does not cool as rapidly at night.

Night-time conditions can be important influences on crop growth. For example, potatoes and sugar beet store more carbohydrates during periods with cold nights, while cotton, maize and tobacco require warm nights for maximum development. Kimball and Brooks[5] have shown how night temperatures are most important for the growth of plants and flowers. In their paper they used the concept of effective day and night temperatures where:

effective night temperature = mean monthly minimum + ¼(mean monthly diurnal range); effective day temperature = mean monthly maximum − ¼(mean monthly diurnal range).

When growing plants are subjected to freezing temperatures, damage or death can result. Certain fruit trees, such as the apple, can withstand extremely low temperatures during winter, while others, such as the fig, would be killed. Many trees have hardy underground structures that manage to maintain life in the more equable soil climate while air temperatures are low, but young seedlings or the flowers can be destroyed.

Frost is of two types, advection and radiation (7.5.4), the effects of which can, in some circumstances, be avoided by protective measures. During frosts, the alternate freezing and thawing can lead to frost heave of the soil or desiccation, both of which can damage or kill the plant. Many plants gradually become acclimatized to the colder weather so that an occasional severe frost in autumn before the plant is adapted can be especially disastrous, as, for example, in the growing of spring wheat. Protection by means of some form of cover or mulch, which reduces the loss of soil heat, can be very effective for root systems of shallow depth. During radiation frosts the crop, such as berries and fruits, can be wiped out while the plant itself is not killed. Many crops, such as potatoes, tomatoes and melons are vulnerable to frosts right up to maturity, whereas some, such as celery and brussels sprouts, appear to benefit from a 'cold snap'.

A universal concept in agricultural meteorology is that of the 'growing season', that is the number of days between the average dates of the last killing frost in spring and the first killing frost of autumn. Such a concept can be misleading because there can exist microclimatic variations from average temperatures over small distances, and because the concept is based on measurements made in the instrument shelter, not in the growing crop.

10.3 Moisture

Water, in all its forms, plays a fundamental role in the growth and production of all crops. The water gain occurs only through the application of water droplets to the soil surrounds or directly to the plant; this may be through rainfall or other forms of precipitation, irrigation or flooding or dew formation. There is no evidence that economic crops extract moisture, via water vapour, from the air, such as occurs with epiphytes. Water loss, on the other hand, normally takes place through transpiration of evaporation, although guttation may sometimes result in appreciable loss.

Soil moisture is the source of water of prime importance to the crop. The supply may range from 'wilting point', when no water is available for plant use, to field capacity, which occurs when saturated soil has lost the water from the large interstices and fissures due to gravity flow. Waterlogged conditions occur when all the soil pores are completely filled with water. A typical heavy soil can hold from 40 cm to 100 cm of water in the top 150 cm. The relative humidity of the soil is not a good indication of the amount of soil moisture because, even with small amounts of soil moisture (2—5 per cent), the relative humidity will still be close to 100 per cent.

When excess soil moisture is present, that is above field capacity, the free movement of oxygen is blocked and compounds toxic to the roots are

formed. The prolongation of such a situation, or even its occurrence, may be due to poor vertical drainage and can often be overcome by correct drainage practices. A soil with a high rate of percolation is unsuitable for cultivation because, when heavy rains occur, the plant nutrients can be removed rapidly. Heavy rain can also damage plants on impact, especially young buds or flowers, and it tends to pack the topsoil layers and delays or prevents the emergence of tender seedlings. The advent of wet weather, or high humidity, can also damage crops or delay their growth: for instance, the drying of wheat and other cereals and the loss of seed and fibre of the cotton plant. A high relative humidity can prove useful in that it will assist in the reduction of transpiration loss but, in some crops, it allows or encourages the formation of mildews or blights, especially within those parts of the crop where air flow is greatly reduced. The formation of such a blight is often made possible by the association of high humidity with some threshold temperature, as in the case of potato blight.

At the other extreme there are the drought conditions, those situations in which the amount of water needed for transpiration and direct evaporation exceeds the amount available in the soil. Unless these conditions are soon negatived by the application of sufficient irrigation water, the plant will begin to wilt and die. Drought may be of three different forms, namely, permanent, seasonal or unexpected. Under the permanent drought conditions of the arid or desert climates, it is appreciated that water must be applied at all times if a crop is to survive. Farmers make allowance for seasonal droughts, such as those that occur with regularity in Mediterranean climates, and irrigation water is, wherever possible, made available. The most treacherous phenomenon is the unexpected drought, for, due to variations in rainfall pattern from the average, a long spell of dry weather can find the crop producer completely unprepared to irrigate. In such a case almost total loss of a crop can occur. To avoid losses of this nature, it is necessary to discover the amount of water (rainfall) needed by a crop at its various stages of development and to ascertain the likelihood that these given amounts will fall at each stage. If he is given this information, a farmer can arrange for protection against economic (cost) considerations. In regions in which the water supply is marginal much can be done to conserve water by suitable management of fallow fields, such as the breaking-up of the surface to reduce runoff and removal of weeds to reduce evapotranspiration losses.

Irrigation is often a lightly used term in regions subject to drought conditions but, as we have seen, too much water can be both physically damaging to the crop and financially damaging to the farmer. The application of the right amount of irrigation, at the right time, is a problem of the agriculturalist and the meteorologist and much research work remains to be done.

Some hydrometeors, such as hail, can cause appreciable damage to crops and many investigations have been directed, especially in Italy, to reducing the incidence of hail by seeding developing clouds (7.5.2). The efficiency of the method is unproven and the farmer normally relies on crop insurance.

Dew can be the source of important, although small, amounts of moisture at certain times of the year in some areas. Some crops respond very well to this extra moisture, tomatoes, peppers, beans, squash and corn for example. It appears that old leaves do not absorb the dew as well as the new. Much work has been carried out in this field in Israel. Dew can affect the spread of some plant disease carriers and fungi that need liquid water for their growth.

10.4 Sunlight

In the biological field many important activities take place due to the presence of rays within the human vision spectrum. For this reason it was decided to include these responses under the heading of sunlight, since the normal understanding and connotation of this word is allied to rays seen by the human eye.

The two main processes are photosynthesis, the basic process of food manufacture in nature, and photoperiodism, the flowering response to daylight.

In photosynthesis the visible rays are the most effective, but ultraviolet rays can influence germination and the energy and quality of seeds. Red light is the most important band in the formation of carbohydrates. For high light intensity the generation of organic matter by plants in the process of photosynthesis is about ten times greater than its expenditure in respiration. The energy associated with photosynthesis is a complex function of the radiation intensity, the ambient temperature, carbon dioxide concentration, etc., and it is found

that, in general, the optimum for normal growth and development is around 8–20 kilolux. Some of the light values giving optimum flowering and fruiting conditions are as follows.[5]

Pea and buckwheat	850–1100 lux
Corn	1 400–1 800
Barley and wheat	1 800–2 000
Bean and cucumber	2 400
Tomatoes and radishes	4 000

If the radiation is insufficient there is a tendency for the stalk to grow at the expense of the foliage while the root system remains under-developed. The average plant begins to accumulate organic matter at about 0 °C, increasing in amount to 25 °C then decreasing to zero at 40 °C. Such figures of temperature were obtained in climatic chambers and are really reflections of radiation as well as temperature. Under fluorescent lights the spectrum is approximately that of normal daylight, but less heat is given off and the uniform distribution encourages photosynthesis.

In the process of photoperiodism the important rays are those between $0.5 \mu m$ and $0.70 \mu m$. Plants are often categorized as being of short-day or long-day types. This means that they achieve their optimum growth or crop maturity time either during the period of short days (around 10 hours' sunlight) or during the long days, around 14 hours of possible sunlight. Some examples of the different types are:[6]

short day — beans, castor beans, corn, cotton, cucumber, millet, sunflower, tomato;
long day — barley, clover, flax, mustard, oats, rye, wheat.

The plants of the tropics are generally short-day types, and plants originating in the mid-latitudes, such as cabbage, dill, lettuce, radish and spinach, have their flowering retarded if the day length is cut — as, for instance, is the case if they are grown nearer the equator.

Some trees also exhibit light preferences. For instance, the birch, larch and pine are light-loving while the beech and spruce are shade-loving.

10.5 Wind

The effect of high winds on crops can be appreciable. Complete physical destruction may result because little can stand against high winds of the order of 100 km/hour; even large trees be-

come uprooted or their trunks are broken. Some
crops have quite a low 'damage' wind-speed. For
instance, the costly 'blow downs' of banana plantations can occur with winds of around 40 km/hour.
In many regions high winds can rip fruit from the
trees, destroy the flowers, or strip the crops.

In Chapter 2 we saw how the evaporation or
evapotranspiration is a function of the wind speed,
so high speeds will lead to an increase of these processes and of the need for more available water.
Strong winds interfere with the activities of insects
during pollination times, but they also assist
directly in the transport of pollen and seeds, including the seeds of undesirable vegetation such as
weeds.

Wind erosion of soil is of prime importance in
some regions of the world where the dry topsoil is
eroded into tiny particles and removed from the
region in the form of dense clouds of dust, such as
occurred in the Oklahoma and Kansas regions during the 'dustbowl' period of the 1930s. When erosion of the soil is taking place there is usually much
drifting, and low growing plants can become
covered and choked by fine deposits.

Some crops suffer abrasions caused by the impact of the tiny hard particles carried by high winds,
a result that can lead either to a spoiling of the crop
or a reduction in its economic value. In some maritime regions salt is carried on the air and can cause
destruction of plants if present in sufficient concentration.

10.6 Evaporation

Evaporation and evapotranspiration have been
shown to be related, provided adequate moisture
is available (3.8.2). In agricultural practice it is
essential to ensure that the plant does not reach
a wilting condition, an effect that can arise if the
soil moisture is at wilting point or if evapotranspiration is taking place so fast that sufficient water
supply for the plant cannot be maintained. Information concerning evaporation should be available so
that the agriculturalist can estimate the approximate
water needs of a crop, otherwise many are damaged
by a lack of water at certain stages of growth
before this condition is apparent by visual
inspection.

The amount of potential evaporation per year,
measured or calculated, can vary greatly from site
to site. Some representative values are:

350–500 mm: cool zone, uniform rainfall, pronounced maritime influence;
750–1 000 mm: cool zone, uniform rainfall, continental influence;
1 500–1 800 mm: Mediterranean climate; cool and
warm zone continental influence; hot zone mountains and islands;
2 500–3 000 mm: warm and hot zone semi-arid
regions;
ca. 3 500 mm: warm and hot zone deserts.

Note that these are only approximate figures,
but they give a good idea of the order of magnitude
of the evaporation losses.

10.7 Phenology

Phenology is the study of natural periodically-
recurring phenomena such as blossoming, migration,
etc., and their relation to climate. For many years
observations recording the dates of occurrence of
certain of these seasonal phenomena have been kept
in many parts of the globe. These have been published in many countries: see, for example, those
of the Japanese Meteorological Agency which include data concerning the use of foot-warmers,
mosquito nets and overcoats, and of the UK Royal
Meteorological Society (no longer issued).

In 1938 Hopkins[7] published an article entitled
'Bioclimatics, a Science of Life and Climate Relations' in which he defined bioclimatics as a science
of relations between life, climate, seasons and geo-
graphic distribution. Thus, he was really giving the
term bioclimatics to the study of phenology. It is
impossible to outline here all the major points
brought out by this interesting and original paper
based on many years of previous investigation. But
in an important sentence Hopkins states '. . . phenomena of life and climate as modified by terrestrial
influences should be equal under equal influences
at the same level across the continents along lines
(isophanes) which depart from the parallels of latitude at the assumed constant rate of $1°$ of latitude
to $5°$ of longitude.' He also adds that a 120 m increase in altitude is equivalent to a $1°$ poleward
shift in latitude. Many of the paper's statements
are so sweeping as to be misleading, but the interested reader is strongly advised to study it because
it contains many examples and unusual concepts.

Although data concerning phenomena and dates
exist, there is often a lack of appreciation of the
important scientific factors or, even if these are

known, they may not be available as measurements. This state of affairs has rather degraded phenology in the eyes of most scientists. However, some aspects of this subject have been investigated. For instance, Nuttonson[8] has investigated the time required from seeding to ripening of the marquis variety of wheat, expressing this purely as a certain number of days. But it must be noted that climatic elements other than temperature play a part in the development of a plant, and the interrelationship and interdependence among the elements constitute a complex problem for research.

The concept of degree-days above a certain threshold temperature value is often used in order to decide the period elapsing between two phenomena. This threshold value varies for different crops, for instance; peas – 4 °C, oats – 6 °C, potatoes – 7 °C, sweet corn – 10 °C, cotton – 17 °C.

Carefully-recorded phenological observations of a crop in a certain area make it possible to derive schedules of interest to agriculturalists, but extrapolation to another area or the finding of analogous regions are problems far from solution. Recently the subject of phenology has found new proponents and may be due for a resurgence of interest.

10.8 Climate and Crop Relationships

Many researchers have attempted to establish relationships between a certain plant response and some simple or complex climatological function. So many of these are available in the literature that it is impossible here to do more than indicate the pattern that has been followed in a few cases.

One of the most popular climatological functions that has been used is the simple one of degree-days above a certain threshold. It must be appreciated though that high temperatures, although contributing to the number of degree days, have an adverse effect upon the growth or production rates of some crops. Recently, work by Brown[9] has attempted to correct this by a study of the relative effects of minimum and maximum temperatures.

Real problems exist in that plant responses to temperature are seldom linear while different combinations such as warm spring–cool summer, or cold spring–warm summer, can give the same value of degree days but very different crop responses.[10]

In 1947 Geslin[11] used an 'action factor' k, where

$$k = t\sqrt{R},$$

t being the mean daily temperature in °C, R being the total radiation on a horizontal surface in cal/cm² day. This factor showed a direct relationship with the leaf growth of grains.

Guyot[12] investigated the grape yield and found it to be related to the excess of the mean annual temperature above 8 °C, while the quality was correlated with a quadratic expression in the hours of sunshine during July.

Hildreth and Burnett[13] found a correlation coefficient of 0·75 between the cotton yield and soil moisture at 1 m depth, measured about the 20th May.

Laude[14] discovered a 0·86 correlation coefficient between wheat yield in Kansas and the rainfall in the period from 1st July to 31st May preceding the crop.

For Hungary, Berenyi[15] has suggested a relationship

$$y = aX_1 - bX_2 + cX_3 - d$$

where y is the best crop yield, X_1, X_2, X_3 are the precipitation, temperature and sunshine during May to July and a, b, c, d are constants. The correlation coefficient was 0·87 — very highly significant.

Prairie wheat production in Canada has been estimated using the conservation of rain and snow on the summer fallow, variation due to major soil types, and the seasonal and areal distribution of precipitation.[16] The coefficient of determination was 0·89, a very significant correlation of 0·94.

In Zambia, Das[17] related the maize yield to rainfall during a 15-day period, the number of rain days during a 30-day spell and the temperature over 20 days to get a correlation of 0·995. A very good summary paper on crop–weather analysis has been presented recently by Baier.[18]

Relationships of the pattern given above usually apply only to a specific crop in a certain climatic zone. The interested reader may find many similar expressions in the comprehensive text of Wang.[19] A climatic classification based on agricultural productivity using the sum of temperatures above 10 °C has also been proposed by Selianinov.[20]

10.9 Artificial Environments

In certain areas attempts are made to grow specific crops in artificial environments. These environments can vary from the simple cold frame to a complicated glasshouse. Some crops are kept in the modified habitat for short periods only and others throughout their growth to maturity and production time. In many cases the method is extremely successful because the natural environment of some crops needs but little modification in order to become optimal. The acres of glasshouse used for growing tomatoes in the Roding valley near London is an example of this.

Perhaps the most detailed survey of the climatology of glasshouses was that carried out by Whittle and Lawrence.[21] Some of their findings are listed below.

(1) A higher transmission of sunlight was obtained in an E–W house than in a N–S house.

(2) Light was most uniformly distributed in an E–W house.

(3) An adequate number of low-level ventilators to complement the roof ventilators is needed.

(4) Winter soil temperatures and heat gain were consistently higher in an E–W house than in an N–S one.

It should be noted that ordinary plain glass will cut down the transmission of ultra-violet rays appreciably, so that bacteria take much longer to be killed.

Jackson[22] has studied the effect of cold frames on air and soil temperature in Kent. He points out the much higher temperatures reached in the frame and notes that within the frame the crop is also preserved from direct wind damage.

Chapter 11

CLIMATE AND FORESTRY

11.1 Introduction

Anyone who has left the open exposure of a field and walked into a forest or dense woodland on a hot summer day has been agreeably surprised by the cooler climate existing there. The dense vegetation canopy keeps out most of the intense radiation, while the moisture fed into the air by the vegetation cools the trunk space to approximately the wet bulb temperature of the ambient air. Within dense woodland the climate is more equable than the neighbouring open terrain, the maximum temperature is reduced and the decrease of radiation to the open sky at night helps to keep the minimum temperature raised. Air movement is greatly lessened and, away from the edge of the stand, very calm conditions usually prevail. Of course, although the forest modifies the climate, the climate modifies the forest. The type and amount of precipitation, the temperatures reached, their diurnal and, more important, their seasonal variation and the soil types all play their part. Also, the forest micro-climate is the habitat climate of the young tree seedlings, and exerts an influence on their rate of growth and sturdiness. It should be pointed out that the cultivation techniques of the forest planners are also most important. Paterson[1] has studied forest productivity around the world and suggests a climate—vegetation—productivity index, but Lauscher[2] questions the validity of this in extreme conditions. The more recent studies of Holdridge[3] utilize the life zone concept to develop the forest productivity index.

Much discussion has appeared in the literature concerning the forest influence upon the precipitation regime of the immediate vicinity. Some say that the presence of a forest thereby increases the amount of moisture fed into the atmosphere and others that the forest reduces the number and intensity of thermals so that convective disturbances are reduced. It is, in the author's opinion, an undecided question at present and the reader is advised to consult Geiger's text.[4] He points out that precipitation formation is a process taking place in the upper atmosphere and that the type of ground beneath is unlikely to have any significant effect upon this process. The works of Mrose[5] have shown that a zone of high humidity exists in the air surrounding a forest, while a paper by Fel'dman[6] states that in the forest zone of the USSR the precipitation amount does not depend on the extent of the forest. Kaulin[7] presents an opposite view, for his investigations showed that the amount of precipitation was higher in the forest than on the adjacent steppe.

Of course, it is true that the presence of the forest greatly modifies or alters the heat and water balance of the adjacent region because both are changed by the soil moisture conditions, the trunk spacing and the crown vegetation density. Also, the time cycle is altered: for example, the water falling on the top leaves of the canopy reaches the soil in a different form, its droplet size spectrum (energy pattern) is altered, the intensity changes and its areal distribution is different.

Schubert[8] in an investigation on a heath in Germany, using a special network of rain gauges, showed that:

(*a*) a 6 per cent precipitation increase could be ascribed to the influence of reforestation;

(*b*) the influence of the forest in dry years was demonstrably greater than in wet years.

It is not possible to make any sweeping generalities about forest climates, for, as noted in Chapter 10, there are many types of forest or woodland, varying from the three- or four-layer vegetation of tropical forest to sparse woodlands, with little or no undergrowth, of the wet and dry tropics. However, the many problems can be tackled if the various patterns and modifications that are known to occur in certain woods are interpreted with common sense when applying the findings to other types of woodland. It must be appreciated that when an area is deforested the creation of a new microclimate may make it most difficult to re-establish a new forest. The practice of leaving one tree, or a small clump of trees, for seed purposes also often leads to failure because the trees cannot survive in their new microclimate, although their physical location is unaltered.

11.2 Radiation

In the normal forest, that is, one in which the crown canopy forms an almost continuous layer, very little direct sunlight gets through to the trunk area or the ground. The actual amount of sunlight depends on the canopy density and the solar angle, but the standard pattern is that small sunlight patches move across the forest floor, each normally heating and lighting a specific place for a short period of time only (Fig. 11.1). Generally,

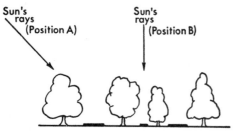

Sun in position A: no direct radiation to forest floor
Sun in position B: direct radiation to areas ▬▬▬

Fig. 11.1 Solar altitude and forest floor radiation

less than 5 per cent of the radiation received by an open field in the same locality reaches the forest floor. Trapp[9] gives an interesting map of the brightness patterns in a beech wood. The heating of the small sunlight areas is intense because of the reduction in, or complete lack of, air flow.

On cloudy days the diffuse, non-directional light

penetrates the interior of the stand and the contrast in radiation between the open and the forest, measured by a light cell, is not remarkable.

Deinhofer and Lauscher[10] have shown that the light intensity estimation of the end of civil twilight (astronomically defined as occurring when the centre of the sun's disc is 6° below the horizon) occurs 16 minutes earlier in a deciduous forest than in the open (this is with cloudless sky conditions), 20 min earlier in evergreen forests and 28 min earlier in an old, high forest. The twilight ended 45 min earlier with cloudy skies, and 54 min earlier in rainy weather. Within the forest the spectral distribution of the light was also altered and the blue light was filtered out much more than the red.

At night the long-wave radiation to the cold sky takes place almost exclusively from the upper, exposed surface and, because there is no heat source similar to that at the ground, the leaves can cool appreciably. Within the forest there is much radiation among the various surfaces but this is small in magnitude because of fairly uniform temperature conditions.

An investigation of how much heat is received on the vertical trunk of a tree has been tackled by Krenn[11] who measured the radiation on horizontal surfaces and introduced geometrical considerations. The method has been extended by Griffiths[12] to apply to slopes of other angles (Fig. 11.2).

11.3 Temperature and Humidity

Temperature is very closely related to radiation. It is the top of a forest which warms first during the day, and sometimes as much as 3 or more hours elapse before the forest floor begins to feel the benefit of the sun's heat. This has been shown to influence the distribution of life within the forest. Similarly, as mentioned above, night-time temperatures are lowest in the upper leaf canopy where the outgoing radiation is greatest. If the night is calm, the cold air from the crown can sink into the trunk space. With a very dense forest this may never happen completely in practice. For all temperature and radiation considerations it is logical and helpful to think of the upper crown as forming the 'effective radiation surface', this three dimensional section taking the part of the two dimensional section of the normal ground surface. Because of the increase in this volume of radiative interplay the temperatures are generally not as high as in open exposures,

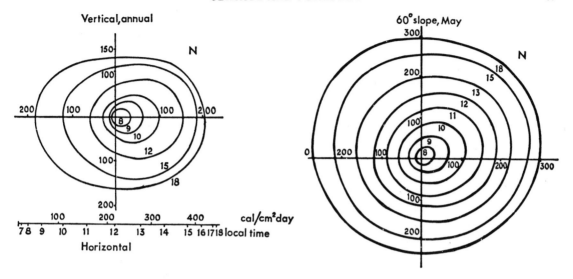

Fig. 11.2 Short-wave radiation incident upon slopes of various inclination, Nairobi

and it appears that the forest has the effect of depressing the mean temperature about 1–2 °C below the annual mean of adjacent areas. For less dense forests these extreme modifications are reduced.

The water output of the leaves, mainly by evapotranspiration within the crown space, principally governs the relative humidity, so that within deciduous forests the relative humidity may often be greater than within coniferous forests. Within the trunk space there is very little air flow and saturation is the order of the day and night. It is interesting to note that most of the investigations concerning humidity measurements in forests are devoted to relative humidity considerations so that pronouncements about dew point and absolute humidity are not possible if long-period observations are desired.

11.4 Wind and Precipitation

A forest acts as a friction obstruction to air flow and a dense forest obstructs in almost the same way as does a small hill range. Very little air penetrates and the air mass is forced to ascend. Within a very close stand the vagaries of air movement are smoothed out in much the same way as radiation temperature and humidity fluctuations are in forest areas.

Geiger[13] gives details of the wind speed gradient, which show the slowing effect of the trees, measured during a period 188 hours in a 15-m high pine stand.

Height of anemo-meter (m)	17	14	11	7	4	1
Speed (km/hour)	5·6	3·3	2·4	2·4	2·4	2·1

Geiger and Amann[14] have shown how the increase of leaf density slows down the air speed within an old oak stand by comparing the number of calm hours with wind speed below 2·4 km/hour.

Height (m)	27 (above crown)	24 (in crown)	20 (low edge)	4 (trunk)
Before leafing (per cent hours calm)	0	8	35	67
After leafing	10	33	86	98

During periods of light drizzle it is only the top leaves that are dampened, and the deposited moisture may evaporate before any appreciable leaf drip can take place. If the precipitation is heavy the crown gets wet, some water flows down the twigs and branches to the trunk and eventually runs down into the soil. Direct dripping through the crowns to the soil is about 70 per cent of precipitation, which means that the danger of soil erosion is reduced when the kinetic energy of the droplets is lost in this way. Erosion is also reduced because the tree root systems have a binding effect on the soil.

If forest trees have large leaves the drops tend to run together and there is a marked gravity force affecting the flow. Sharp needle-type leaves cause the drops to fall directly to the ground.

A phenomenon that has not been fully investigated is that of horizontal or fog precipitation Moisture-laden air, such as occurs in fogs or low stratus clouds, meets with surfaces such as leaves, branches, rocks, and buildings, which then act as collectors and concentrators of the suspended water droplets. This can happen on the edges and crowns of forests and investigations by Nagel[15] on Table Mountain, South Africa, have shown that the collected 'precipitation' may equal in amount that due to normal precipitation.

11.5 Stand Patterns of Climate

Pronouncements concerning the variation of climatic elements in stands of forest cannot be made here because the types of stand are legion. Generally, however, variation is less under a dense crown or undergrowth and larger fluctuations occur under a light cover.

Openings in the forests present a special problem and may become veritable hot spots if they are of the right dimensions. If the size is such that the sun's radiation can penetrate freely, yet the air flow is sufficiently reduced, the volume of air held in the clearing will be heated appreciably, especially if the ground surface is covered only by short or sparse vegetation. Then, similarly, the clearing can turn into a 'frost hollow', or cold pool, at night when radiative cooling is at a maximum.

If the clearing is of a diameter greater than about three times the height of the surrounding trees there can exist an induced air flow within the clearing. This occurs when the air at or just above the crowns reaches the clear spot, where friction is suddenly reduced, and begins to assume a definite downward component. During the daylight hours the added vertical currents of the thermals can cause an almost independent cell to develop, and at ground level the air may be flowing at 180° to that in the free air above the forest (Fig. 11.3).

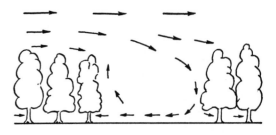

Fig. 11.3 Schematic wind flow in a forest clearing

11.6 Timber Production and Forest Fires

Dense stands tend to develop tall, straight trees of small diameter as the trees fight upwards for light. Knowledge of the climatic environment which is best suited to growth at all stages of the tree development is basic to efficient management for timber production, watershed protection and other forest economics. As trees have such a long life they are influenced more by the climatic fluctuations through the years than the vagaries of day-to-day weather. There are, of course, exceptions to this rule: for instance, the occurrence of intense, disturbances, such as tornadoes and hurricanes, which destroy individual trees or whole forests.

Griffith[16] has shown that by using a running balance of soil moisture that is calculated from easily measurable climatic elements, it is possible to increase the chances of survival of exotic softwoods during early growth in an area with an unreliable climate.

Fritts[17] has studied the relationship of radial growth of beech, white oak and sugar maple to maximum and minimum temperature. He found that for maximum temperature up to about 21–27 °C the growth increases, but decreases from then onwards. He tested the growth rate against seventeen variables but, due to limitations of statistics, it is always possible to overlook an important significant correlation if it is of a polynomial, implicit nature. Such relationships can be tested only if there exist other non-mathematical sources of information.

Keen[18] has shown that the Oregon ponderosa pine has a growth rate related to the 2-year cumulative departure of precipitation from the mean, illustrating how soil moisture acts as a balance or integrator over short periods.

Insect pests, wood borers and the dispersal of seedlings are greatly affected by the climatic and weather conditions. This is an aspect of the subject that has been investigated little so far. Curry and Griffiths,[19] studying the incidence of *oemidae* in tropical trees, measured the temperature reached within dry and wet logs of cypress and found that at 1 cm within a dry log exposed to the sun the temperature was 54 °C, while at 10 cm it was about 43 °C. Such an investigation shows clearly how very different is the micro-environment of the animals living within the trees themselves; and that to attempt to find correlations between observed behaviour and conditions external to the actual habitat is a sheer waste of time. It is necessary first to measure the environment of the organism.

Dispersal of seeds is aided by primary and secondary air flows. But some thermals can play an important role, such as those in clearings (11.5) or ones that are sufficiently developed to entrain the seeds into the free air flow.

Forest fires are perhaps the most obvious economic manifestation of the importance of climatic or weather conditions. The weather is all important because of the part it plays in determining whether the undergrowth is in a dry state, while the factors of relative humidity, temperature, wind speed and direction, precipitation and the condition of the vegetation and litter are fundamental. Many forestry organizations use a simple fuel moisture indicator made of wooden sticks mounted in the open and freely exposed to the weather, to yield information about the relative danger of fires. The sticks, which lose their moisture in the drying conditions in which forest fires are likely to occur, are weighed and the danger is assessed against their moisture content.

The weather bureaux often give special forecasts for the areas subject to forest fires, stressing the elements important in this problem. In the USA use is made of the Forest Fire Danger Meter, an indicator that assesses the integrated value of such factors as days since the last rain, the amount of the last rain, wind speed and fuel moisture. There is a possibility that cloud-seeding (7.5.2) may help to control fire outbreaks in some areas.

11.7 Windbreaks

Windbreaks or shelterbelts are generally a single, occasionally multiple, row of trees planted to afford some protection to specific crops. It is known that such a row of trees is instrumental in reducing the air speed in its lee and, under certain conditions, this may be desirable or essential in order to prevent the destruction of flowers or fruit. For a fuller discussion of this subject see 7.5.1, Read,[20] or the W.M.O. report.[21]

Chapter 12

CLIMATE AND HUMANS

12.1 Introduction

Every day man, wherever he may be on the earth's surface, has to live with the weather. In the regions of the world where weather exhibits day-to-day changes it is general to open a conversation with a few comments about the weather, with the result that we are made more and more aware of the elements and the part they play in our lives.

In the more developed countries the 'weather forecast' is listened to avidly, for it helps one to decide just how to dress in order to be at ease with the weather. In less developed areas, where such a service may not be available, man just awaits the occurrence of changes and then does his best to adapt to them.

The effect of climate on the progress of history and the development of civilizations is called climatic determinism. Huntington[1] developed this theory while Markham[2] expounded a concept relating climate and the energy of nations, substantiating his idea with ancient and modern examples.

In the early days of mankind it was in the regions where man did not have to put up a continual fight with the elements in order to survive that the first civilizations developed. As a small human community did not have to worry unduly about clothing and heating problems it was able to devote its time to 'the sword or the pen', if an anachronism is permitted.

The nude, semi-reclining male needs a temperature of 29 °C for comfort, whereas, clothed, he is at ease at 25 °C.[3] We may therefore suppose that, as man developed clothing and heating methods, he was able to move into less climatically-favourable areas and still have time to concentrate on the more important things of life. With the improvement of the insulating power of clothing, better heating facilities and the building of homes suitable for the climate, man could combat the elements so that the developed civilizations of those times could move into previously inhospitable regions. It is a fact of human physiology that the human can perform physical activity better at a temperature above the optimum for mental labour and so, with the stimulus of a colder climate that his body could now easily accept, mental development seemed to accelerate in cooler regions with marine or semi-marine climate. The rigours of a continental climate still appeared to be too much to combat. Manley[4] has reviewed recent studies in this aspect and Rigg[5] has presented an interesting series of articles in which he discusses the declines of both the Greek and Roman civilizations in the light of possible plagues. He concludes, rightly, that there is need for more scientifically based study. Nowadays, of course, man can take with him almost anywhere just the climate to which he is accustomed, providing that money is available to meet the cost, so it is likely that the people of most areas of the globe will begin to develop mentally at an appreciable rate when economic standards increase. It must be borne in mind, though, that man can alter the climate of small volumes only and much research work still must be undertaken in enervating or raw atmospheric conditions.

12.2 Heat Balance

In order to obtain some idea of the various complex factors in the heat balance of man it is best to consider him as an engine, needing fuel to exist. Under this premise it can be said that the average man (weighing 70 kg with a surface area of about 2 m^2) when completely at rest (sleeping) will give out 80 kcal/hour, an amount which, if not dissipated, would serve to raise the body temperature by about 1 °C/hour. When walking at about 5 km/hour this amount increases to over 200 kcal/hour while, under optimum conditions, with extreme work it can be increased to about 600 kcal/hour for 1 hour or so. If we are dealing with a man of other than 70 kg weight these figures should be altered proportionally to (weight)$^{0.67}$. Metabolic heat production is increased if the man is carrying a pack in addition to his own weight, the increase being about 3 kcal/hour/kg for small loads and about 4 kcal/hour/kg for loads up to 20 kg. The main source of body heat is food, about 80 per cent of which is employed in growth, body repair and heat production, while 20 per cent remains as energy for daily activities. During muscular activity about 70 per cent of the heat produced is waste and must be lost or dissipated.

In addition to this metabolic heat, man in a natural environment can gain heat from radiation, convection and conduction. The schematic pattern is shown in Fig. 12.1, where the heat sources are depicted to the left and the heat sink losses, to the right.

Man can receive a radiative heat load from the environment provided that there is some radiative surface, in direct line with some part of his body, having a temperature greater than 33 °C, the approximate average body surface or skin temperature. Adolph[6] has shown that, to a first approximation, this gain can be expressed in kcal/hour as:

$200 + 25(T - 33)$ for a nude man in the sun, reasonably independent of his pose;

$100 + 22(T - 33)$ for a clothed man in the sun;

$20 + 18(T - 33)$ for a clothed man at night;

where T is the air temperature in °C.

These figures are based upon experiments carried out in desert (arid) areas. In regions of high humidity they will be different and would show a reduction of the radiation heat load. It is interesting to note the great reduction in radiation experienced by the clothed man as compared to the nude man in desert areas.

So as to obtain an integrated view of all the surfaces that are playing a part in the radiative balance, the concept of a mean radiant temperature has been introduced. This is simply the temperature at which an object gives out as much radiation as it receives from its surroundings, termed the radiative equilibrium temperature. An estimate of the mean radiant temperature can be made using the globe thermometer, a copper sphere with a thermometer held in the central air space.[7]

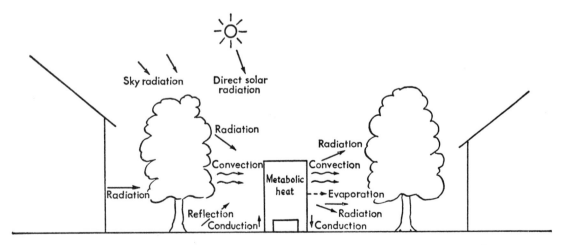

Fig. 12.1 Schematic heat balance for an animal in a natural environment

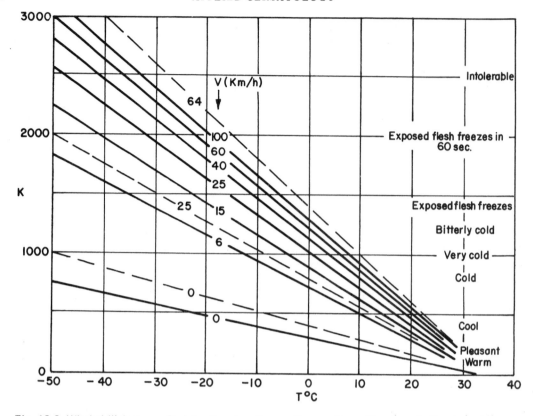

Fig. 12.2 Wind-chill factor — Siple's values are the continuous lines, Steadman's the dashed lines

Man will, of course, gain extra heat when the ambient air temperature is above 33 °C and air movement advects more heat to the body. However, under conditions in which the air temperature is less than 33 °C, a wind-chill effect will occur. This convectional heat loss is greatly increased when the temperature is low and the wind speed is high. Siple[8] has studied the cooling power of the air in motion and has proposed the use of the 'wind-chill' factor. This is based on the rate at which the naked body would cool, the factor being altered radically by the presence of clothing. The sense or feeling of cold by the hands and the face does, however, check the factor fairly well.

The value of the wind-chill factor,[9] K, is given, in kcal/m^2 hour, by the expression

$$K = (33 - T)(10\sqrt{v} + 10 \cdot 5 - v),$$

T in °C, v in m/s.

The variation of K with T and v is shown in Fig. 12.2.

The sensation scale for K is as follows.

50	hot	100	warm
200	pleasant	400	cool
600	very cool	800	cold
1 000	very cold	1 200	bitterly cold
1 400	exposed flesh freezes		
2 000	exposed flesh freezes in 60 s		
2 500	intolerable		

It should be noted that K reaches a value of 1 400 under any of the following sets of conditions.

$$-7\ °C \text{ and } 70 \text{ km/h}$$
$$-12\ °C \text{ and } 30 \text{ km/h}$$
$$-23\ °C \text{ and } 11 \text{ km/h}$$
$$-40\ °C \text{ and } 13 \text{ km/h}$$

Such a table indicates how important is the role played by air flow at low temperatures and shows how man can be just as uncomfortable in the maritime climates during high winds as in the continental climates which generally have relatively calm conditions during the winter. Recently Steadman[10] has improved the concept by using a method that applies to persons clothed adequately to maintain thermal equilibrium. The differences between the two models are seen in Fig. 12.2.

The conduction of heat to and from man is usually quite small, taking place through the feet or foot clothing, but it can assume important proportions when a man is resting on the ground, especially at night. In this case much heat can be conducted from the body to the cold surface.

So far no mention has been made of the role of water in the thermal regulation of man. A non-evaporating body will only achieve equilibrium at a temperature just balancing all the heat source loads and sink losses. Note that this will not necessarily be equal to the mean radiant temperature. Man, however, if he is to survive, must keep his body temperature within strict, small limits; and under conditions of heat stress the evaporative heat loss becomes dominant.

12.3 Water Balance

When a gram (1 cm^3) of water is evaporated completely 0·58 kcal of heat are used. Expressed differently, it would take about 136 kcal to evaporate a cup (230 g) of water. When the body begins to get overheated it has recourse to the simple expedient of losing heat by the evaporation of water, either by sweating or by direct evaporation of moisture from the lungs and upper respiratory tract. At temperatures above 30 °C and with high humidity (>50 per cent) the 'breathing' loss is much less than the 'sweating loss. If the air is saturated and at a temperature above 33 °C, only the latter method is available. When the temperature is above 37 °C (deep body temperature) even this possibility is denied to man. Expired air has a relative humidity of about 80–90 per cent[11] so that respiratory heat loss is not maximal.

When this set of climatic conditions occurs artificial means of alleviation must be applied, or heat prostration and death will follow since a rise in the body temperature of a few degrees can cause permanent and irreversible damage to the brain cells.

The average human body contains about $\frac{2}{3}$ its weight in water, yet, even with such an appreciable quantity, a deviation of as little as ±1 per cent can cause physical disturbance, while with a loss of 10 per cent man cannot walk and at 20 per cent he is unable to survive unless he is rehydrated in a reasonable time.

Adolph,[6] again for desert conditions, has found the following sweating rates in g/hour for an average man.

$720 + 41(T - 33)$	walking in the sun
$400 + 39(T - 33)$	walking in the night
$300 + 36(T - 33)$	sitting clothed in the sun
$180 + 25(T - 33)$	sitting clothed in daytime shade

Note that from the previous equations (13.2) clothing saves about 120 kcal/hour ($T = 39$ °C), an amount equivalent to about 200 g/hour.

It is necessary to assume that there is sufficient air flow for the saturated air in contact with the moistened body to be carried away. This is generally the case, but it is to the detriment of the body's water balance if this air speed is higher than necessary, for too much water loss ensues. It is *not* a good idea to move around in order to create a breeze because the metabolic increase would then offset any air flow advantage.

If man is to lose water he must have an intake source because it is essential that water lost by sweating or respiratory evaporation soon be replaced. Water loss fatigue is an insidious thing, and men can collapse from a water deficiency without realizing the cause. There are cases on record or men collapsing while still having full water containers in their possession. Thus, a knowledge of the above values, and their meaning, should prove of use to anyone who finds himself under conditions of heat stress, even when on a hike on a warm summer day.

12.4 A Balance of the Body

As a summary of the previous two sections it is clear that the body gains heat from

(a) radiation from surfaces at above 33 °C (such as sun, lights, etc.), R_+;

(b) convection, the warmth from the air when temperature is above 33 °C, C_+;

(c) conduction, from physical contact with objects at a high temperature, P_+;

(d) metabolism, ranging from basal to about ten times basal, M.

The body can lose heat by

(a) radiation, to surfaces at a temperature below 33 °C, R_-;

(b) convection, by air flow and natural heat thermals from body, C_-;

(c) conduction, from physical contact with objects at lower temperature, P_-;

(d) evaporative loss, E.

If the body is in thermal equilibrium, which is naturally desirable,

$$R_+ + C_+ + P_+ + M = R_- + C_- + P_- + E.$$

Normally, for the nude person, except when lying in contact with hot ground or with wet, cold ground, P_+ and P_- are small. But for the clothed human conduction with the air layer trapped in the clothing can, under some circumstances, be large.

It is interesting to note that for temperatures less than about 10 °C, $R_- + C_- = 9E$, while at 21 °C, $R_- + C_- = 4E$.

E at these temperatures is in the form of insensible loss, mainly through the lungs. The relationships are, naturally, only approximate. At 30 °C $(R_- + C_-)$ and E are about equal and at higher temperatures E begins to dominate, until at about 33 °C the radiative and convective losses are nearly zero.

12.5 Empirical Estimates of Physical Feeling

A number of investigators have endeavoured to express the effect of climatic parameters upon human comfort by means of certain expressions embodying two or more of these parameters. A few of the more generally used expressions follow.

12.5.1 *The Effective Temperature*[12]

One of the earliest indices used is the effective temperature. In this case the existing conditions of temperature, humidity and wind speed are related to still, saturated air at a new temperature, E.T. The value of E.T. is generally derived from a graph but, to a first approximation, the E.T. can be equated to the T.H.I. (12.5.2), especially over the range of 16–32 °C). It is suggested in the literature that an E.T. in excess of about 31 °C is a stressful condition for the body, with an E.T. of 35 °C being a good upper limit of tolerance.

12.5.2 *The temperature–humidity index*

This index, originally called the discomfort index, is meant to express the degree of discomfort felt by the average office worker, hence the aspects of radiation and wind flow are not considered. As a measure of the temperature–humidity index under such circumstances, Thom[13] has suggested the use of the expression

$$\{0{\cdot}72\,(T + T_d) + 41\}$$

or

$$\{T - (1 - 0{\cdot}01\ \text{r.h.})\,(T - 14{\cdot}5)\},$$

where T is the air temperature, T_d is the dew point temperature and r.h. is the relative humidity. This expression is just the effective temperature (12.5.1) obtained by a simplified method. Values of the temperature–humidity index for representative temperatures and relative humidities are given in Fig. 12.3.

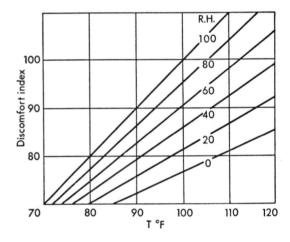

Fig. 12.3 Variation of discomfort index with temperature and relative humidity

From the reactions of a number of people, Thom has found that for a temperature–humidity index below 21 no discomfort is felt, between 21 and 24 some people are uncomfortable, 50 per cent are uncomfortable at 24, and above 27 most people are showing signs of discomfort. For values above 29 there is distinct stress and some US Government Offices dismiss their employees at such times. A value of 33 was noted for Yuma,

Arizona, in July 1957. This form of the equation is empirical and is based on the sensations of a large number of people of similar culture and reaction to environment, so that it may not apply to, say, an African or an Asian.

12.5.3 *The strain index*

There have been numerous attempts to measure the relative strain (R.S.) imposed upon man by a certain set of atmospheric conditions. The best index so far is that developed by Lee and Henschel,[14] which includes the metabolic rate, air temperature t_a (°C), humidity p_a (mm Hg), work rate, resistivity of air and clothing to outward flow of heat and the passage of water vapour, and the volume of expired air. A simplified form, using the metabolic rate of a person walking at 3·2 km/hour, with a lightweight business suit in a breeze of 0·5 m/s is

$$\text{R.S.} = 10·7 + 0·74 \, (t_a - 35) \, (44 - p_a)^{-1}.$$

Other comfort indices have been proposed and the interested reader is referred to a comprehensive summary by Landsberg.[15]

12.6 Clothing Insulation and Clothing Zones of the World

Siple[16] has given details of a classification of the world according to clothing requirements based, to a large degree, upon the climate. In this he distinguishes seven main zones while making the point that because of the different metabolic rates of active or inactive persons, it is difficult to make sweeping general statements as to the exact clothing needed.

(1) *Minimum clothing zone – humid tropical and jungle type.* A nude man at basal metabolism will begin to feel cool at air temperatures below 30 °C. When standing still he can easily endure temperatures down to 29 °C, while, if marching temperatures as low as 24 °C can be accepted. These figures apply to males of North American stock only, for it is known that by adaptation people can endure much lower temperatures while completely nude, enduring them for long or lifetime periods, as do the Australian aborigines or the Onas of Tierra del Fuego. If a small covering is worn over the torso these temperatures can be reduced by several degrees. The minimum clothing

zone comprises the areas where the mean monthly temperature varies between 20 and 30 °C. Here the primary factors are:

(*a*) adequate clothing to satisfy traditional modesty and fashion;

(*b*) protection from solar rays for skin and eyes;

(*c*) protection from pathogenic organisms, thorns and insects;

(*d*) camouflage, when this is necessary.

The suggested covering is light cotton or similar material.

(2) *Hot dry clothing zone – desert type.* Here are the areas of warm air (often above 33 °C) and high radiation. Clothing is needed for protection from the intense solar radiation, but it must allow evaporative cooling and provide insulation against the strong night-time cooling. It is a region of little rainfall. For this zone long flowing robes are advised. Although light-coloured clothing will absorb less radiation than dark clothing, the difference is not so great as may be expected since dirt particles and the rough, matt nature of materials soon make most clothing an effective 'black body'.

(3) *One-layer clothing zone – subtropical or optimum comfort type.* Here, where the mean monthly temperatures vary between 10 and 20 °C, the torso needs protection, but not the extremities. The clothing advised is wool or similar material with light cotton undergarments. In the humid (wet) parts of this zone there is the need to protect from the cool rain and, while a thin waterproof is satisfactory, it also reduces convective loss and causes heat storage; an umbrella is the best solution.

(4) *Two-layer clothing zone – temperate, cool winter type.* In this zone the weather is frequently humid with occasional snowfalls. The mean monthly temperatures vary between 0 and 10 °C and there is not an excess of radiation. The clothing must insulate against the conductive heat loss and reduce air movement yet allow the passage of insensible perspiration in the vapour state. External wetting must be excluded. The ideal type of clothing depends very much on the state of activity but it should, for inactive occasions, allow a layer of air of about 6 mm to be trapped between the two layers of clothing but permit

the ready removal of the outer garments when manual labour is undertaken.

(5) *Three-layer clothing zone – temperate, cold winter type.* Mean monthly temperatures between −10 and 0 °C; snow and ice are prevalent. Exertion when wearing the three layers, with two trapped layers of insulating air, causes sweating so that the clothes must be openable to allow loss of heat. This means that thick pullovers are not a good solution.

(6) *Four-layer maximum clothing zone – sub-arctic winter type.* Siple concludes that, for this zone, no single system of clothing can be said to be completely superior to all others. The Eskimo fur clothing is perhaps the most efficient from the standpoint of weight, utility and simplicity, but it is costly. No clothing suitable for an actively working man can provide protection in extremely cold weather for more than a short period, for instance when the man stops work to rest or to stand around or to sit in a conveyance. Temperatures here range from −20 to −10 °C.

(7) *The activity balance zone – Arctic winter type.* This represents the extreme of cold stress, conditions under which comfortable heat balance cannot be maintained by any simple increase of clothing insulation alone.

In considering the problems of clothing and general living in a new climatic area, it must be appreciated that a degree of acclimatization or adaption to the fresh environment takes place in the body, but Siple also stresses the accustomization factor. This term is meant to imply that the individual must learn how to tackle sensibly the act of living in the new environment, to learn from others more experienced, to apply common sense to actions and dress and, unfortunately, 'to learn from and by mistakes'.

Under conditions of hot or cold stress it is necessary for the body to be insulated, or at least partly cut off, from the source of thermal trouble. In general, this is effected, for man, by the wearing of clothing. However, it should be noted that even the air alone has an insulating effect. This effect can be expressed as follows.

Insulation of air
$$= [0 \cdot 61 t^3 + 0 \cdot 19 t^{-1} \cdot \sqrt{v} \cdot \sqrt{(p/p_0)}]^{-1}$$
where the insulation is in clo units, $t = $ (air temperature, K)/298, $v = $ wind speed in cm/s, and

$p, p_0 = $ atmospheric pressure at station level and sea level respectively.

The clo is a unit of insulation commonly used in clothing studies, equal to $0 \cdot 18$ °C/kcal/m²/hour It can also be thought of as the insulation needed to maintain the skin at a temperature of 33 °C with a room temperature of 21 °C, an air speed of less than 3 m/min and a relative humidity of below 50 per cent, when the metabolic rate is 1 Met, or 50 kcal/m²/hour. It is approximately the insulation of a 'business man's' clothing.

A basic equation developed from the laws of heat flow shows that, if H is the total body heat loss, in non-evaporative conditions,

$$H = (T_s - T_a)/(I_a + I_{cl})$$

where H is in kcal/m² hour. T_s and T_a are the skin and air temperatures in °C, I_a, I_{cl} are the insulation of the air and clothing in °C/kcal/m² hour. If we allow that, for comfort,

$$T_s = 33, \quad \text{then} \quad T_a = 33 - H(I_a + I_{cl}). \quad (1)$$

Heat loss must balance heat gain if constant body temperature is maintained and it has been suggested that, allowing for evaporation and work dissipation,

$$H = 0 \cdot 75 M, \quad (2)$$

where $M = $ heat production. At low values of $H(40)$ and 1·5 km/hour wind, equation (1) gives $T_a = 29$ °C, for a nude body. For a clothed person ($I_{cl} = 0 \cdot 18$, say), $T_a = 22$ °C. If a person is doing light work ($H = 80$), $T_a = 26$ °C (nude), 11 °C (light clothes).

It is a known fact that wind blowing on exposed skin will cause a cooling effect. This thermal wind-decrement, T_{wd}, has been shown to be approximated by $9HW$ (°C), where H is in Mets and W (clo) is the reduction in the insulation of the air due to the wind speed. For 1·5 km/hour, $W = 0 \cdot 5$, and at 80 km/hour, $W = 0 \cdot 9$. Note that this decrement is a function of the metabolic rate, increasing with greater metabolism. The decrement can be used to adjust T_a in the basic equation so that, with wind effects,

$$H = (T_s - T_a + T_{wd})/(I_a + I_{cl}).$$

An important application of this approach is in the case of very cold temperatures. For instance, if T_a is −17 °C then exposed flesh must have heat brought to it at a rate, h, if it is to maintain constant temperature, where

$$h = [33 - (-17)] / 0.18 = 275 = 5.5 \text{ Mets.}$$

Thus, with an 80 km/hour wind the thermal wind-decrement is

$$9 \times 5.5 \times 0.9 = 45 \,^{\circ}\text{C}$$

This makes conditions intolerable.

A similar approach can be used to account for the thermal increment due to radiation. This increment, T_r, is given by $9RI_a$ where R is the absorbed radiation in Mets. Reverting back to the figures given by Adolph, we can assume, for hot desert conditions, R is about 3.5, and with a 1.5 km/hour wind T_r is 16 °C. If the derived equation is written in another form it can be said that, to preserve the previous equality when incident radiation R is experienced, the heat loss must be increased by $RI_a (I_a + I_{cl})^{-1}$. The term $I_a (I_a + I_{cl})^{-1}$ is the efficiency factor, which is zero for the nude human and about $\frac{1}{3}$ for a person in light clothes in a 1.5 km/hour wind.

The maximum practical insulation of clothing is about 1.5 clo/cm thickness. But it must be realized that the insulation of fur, clothing and other protective coverings is dependent upon weather conditions. For instance, during rain or high humidity conditions, the insulation of wool is reduced because its air spaces become partly water-filled and reduced in size.

12.7 Climate and the Home

Climate has a most subtle way of influencing both our daily lives and way of living. The weather can be our ally or our enemy, and this fact is particularly true when considering the influences that climate has upon household management.

12.7.1 Food

Food, or its method of preparation or cooking, is affected by climate through variations in temperature, humidity, sunlight and pressure. Dietary habits are similarly influenced by climatic conditions. 'Climate' in this section refers of course to the outside weather conditions, although it is fully realized that different, sometimes artificial, climate may prevail indoors.

Temperature. A rise in temperature will cause increased enzyme and micro-organism activity which results in the rapid spoilage of foodstuffs unless some precautions are taken for their preservation. Enzyme activity speeds up the decay of fruits and vegetables.

Moulds are actively reproduced under warm, humid conditions. Fruit, cheese and bread are particularly susceptible to mould growth. Most varieties of mould are killed at 70 °C, with the exception of *Byssochlamys fulva*, a rare mould of fruit which is destroyed at 85 °C.

Yeasts have an optimum growth around 15–20 °C and fermentation in such processes as wine- and bread-making is accelerated under these conditions. Generally bacterial activity and growth are optimum about 37 °C, blood heat.

Where ideal conditions of warmth, moisture and a suitable food prevail bacteria reproduce and cause rapid decay of foodstuffs, in particular perishable foods such as milk, meat, fish, eggs, fruit and vegetables. Serious food poisoning can be a direct result of eating foods which have been contaminated in this way. Some bacteria, however, perform an essential function in cheese-making and ripening, and also in beer-brewing and wine-making.

An increase of heat also changes the composition of fats, causing them to liquefy, the melting point varying with the type of fat. It is for this reason that pastry-making can become a problem in hot weather.

Evaporation of the liquid content of foodstuffs increases with temperature rise, providing that the humidity remains low. This causes bread, cakes, etc., to become stale very rapidly if unprotected by a tin or air-tight covering. The drying of foodstuffs by evaporation is one of the oldest methods of preservation still being practised out of doors in some regions of the world. Examples are tea, coffee beans, fruits and fish. It is particularly successful where adequate air flow is combined with hot, dry conditions.

A decrease in temperature has the reverse effects from those of temperature rises. Micro-organic growth rate is slowed, if not completely stopped and yeast growth is inhibited below 10 °C. Moulds, although they tolerate temperatures as low as −10 °C, do not grow. Many types of bacteria are destroyed by extremely cold temperatures, but others will simply lie dormant in a spore stage at below-freezing conditions. Fats, and even oils, solidify at low temperatures.

Humidity. High humidity in the atmosphere affects several normally dry foods unless they are adequately protected. Cereal foods, for example wheat flour and oatmeal, may deteriorate rapidly; sugar, salt and powdered gelatine being hygroscopic, absorb moisture and become lumpy. High humidity, together with a rise in temperature, causes an increase in micro-organism activity. Low humidity is often utilized in hot climates to induce excessive evaporative loss to water kept in porous pots. This loss of heat by evaporation keeps the water at about wet bulb temperature.

Sunlight. Sunlight destroys vitamin C, particularly when the vitamin is contained in a liquid form in jars or bottles which permit the penetration of destructive rays. Milk, fruit juices and bottled fruit can be similarly affected. Excessive sunlight is also responsible for the wilting and yellowing of green vegetables.

Pressure. At sea level the standard atmospheric pressure is about 1 kg/cm^2, but this decreases at a rate of about 30 gm/cm^2 for every 300 m in elevation, up to a height of some 3 000 m. Day-to-day weather conditions are responsible for slight pressure variations, but these cause no great problems to the cook.

An appreciable decrease in atmospheric pressure, however, does cause problems with raising mixtures and boiling. The standard cookery recipes do not, in general, need any adjustment for use up to elevations of 1 000 m but, as elevation increases above that level, the problems become more acute.

Flour mixtures need special attention if they are raised by carbon dioxide, air or steam since expansion of the gas is greater where the air pressure is reduced. A few specialized cookery books[17] give some of the methods used to overcome these problems — for example, a reduction in the amount of baking powder or yeast used in the recipe.

Liquids boil at a lower temperature at higher altitudes. For pure water the decrease is about 1°C for each 300 m increase in elevation. A pressure cooker may be used to speed the process but steam within the cooker is also affected by the altitude. To reach the required temperature, for example when canning, so as to ensure complete destruction of harmful micro-organisms, the pressure setting of the cooker must be increased by about 0·1 kg/cm^2 for each 1 000 m elevation.

When sugar solutions are boiled, in making sweets, icing, etc., it must be remembered that the boiling temperature is also lower. It is agreed by connoisseurs that this lower boiling point of water can affect the making of a good pot of tea.

Dietary habits and cooking methods. Climatic variations have a direct bearing upon the type of foods we eat, though this is possibly less true today than it has been in the past history of mankind.

In cold climates there is a tendency for people to eat more fat and carbohydrate foods to meet the higher calorific needs of the body. For example, Eskimos have a high fat intake of oil obtained from seal, walrus or whale blubber. In cold climates, too, cooking is done inside the homes, the stove often performing the dual function of a heating appliance and a cooker.

Hot climates influence food habits in that uncooked foods such as fruit, vegetables, cold dishes and others with a reduced fat and carbohydrate content occur more frequently in the diet. There is also a considerable increase in liquid intake to make up for the loss through perspiration. Outdoor cooking and food preparation is frequently seen in a hot climate.

12.7.2 *Clothing*

The direct effect of climate on the types of clothing worn has been dealt with in 12.6. This section discusses the effects of climate on fabrics used in making clothing.

Humidity. A high humidity makes the outdoor drying of laundry difficult, and provides ideal conditions for the growth of mould which affects leather goods and garments made of vegetable fibres (mildew). It can also cause the deterioration of covers of leather-bound books.

Cotton and linen garments in conditions of high humidity tend to become limp and crease readily because of slight moisture absorption while wool garments soon become damp due to their highly hygroscopic nature. Finished wool materials can contain approximately 12–14 per cent of water.

In especially damp climates it is a good thing to have some form of heater in the clothes storage cupboard because this keeps the air at a high temperature and reduces the relative humidity.

Wind and sunshine. Moderate winds and sunshine provide ideal drying conditions for laundry; strong winds, however, tend to force garments out of shape.

Sunshine has, for centuries, been utilized for the bleaching of linen cloth whch is spread out of doors, but it unfortunately tends to make white woollen cloth turn yellow. Prolonged sunshine causes some dyed fabrics to fade, and accelerates the rotting of rayon and silk fabrics in particular. These facts should therefore be borne in mind when choosing furnishing fabrics, especially curtains. Venetian blinds are most useful in protecting furnishings from direct sunlight, since they can easily be adjusted to shield them from direct rays but still admit normal daylight.

Rain. Rain water, since it is extremely soft, having few dissolved minerals present, can be an asset to laundering garments, particularly woollens.

12.7.3 *Cleaning*

Each location, with its particular climatic pattern, poses its own cleaning problems. For example, a dirty industrial atmosphere with a predominantly high humidity presents very different problems from those of a clean country area with a dry climate.

Humidity. High humidity often causes condensation on such surfaces as polished wood, metals and mirrors, and there is little point in polishing them until the atmosphere becomes drier. Metals tarnish readily in these conditions; copper, for instance, becomes covered with a basic carbonate called verdigris.

Damp walls result in the staining and displacement of wallpapers and even plaster. Wooden doors, window frames, etc., have a tendency to swell and warp unless they have been well seasoned and preserved.

Fog, with its 100 per cent relative humidity, is a prime cause of the conditions discussed above, and smog, which is a combination of fog with the dirt particles of an industrial atmosphere, introduces added cleaning difficulties because of its sulphur and oil deposits.

Sunshine. Dry, sunny days are ideal times for carpet cleaning, whether for drying shampooed rugs or for drying the dust in them so that it can be readily removed by shaking and beating. Unseasoned and unprotected woodwork dries and may even split if exposed to sunlight, and paint dries and flakes — although improved qualities which can withstand extreme weather conditions are now on the market.

12.7.4 *Gardening*

All aspects of gardening are affected by climate, but the following examples serve to illustrate how climatic conditions may assist or deter gardening processes.

Warm drizzly days are ideal for bedding out plants. Weeding is made easier after a shower of rain because it loosens the soil, and it is preferable to fertilize plants with a dry fertilizer before a heavy rainfall so that the plant nutrients may be washed well into the soil. It is inadvisable to water plants in the heat of the day, however, because excessive evaporation results and plants are likely to become chilled by the loss of the heat used in vaporization. If watering takes place late in the evening the high humidity induced during the night can lead to blight and fungus problems.

Calm, dry weather is the best time for spraying, while dry days should be chosen for mowing. If high winds are forecast it is obviously advisable to collect maturing fruit crops in advance.

Soil which has previously been dug benefits from the advent of frost since this assists in the soil break-up. This is why gardens are traditionally dug in the autumn.

A useful book dealing with gardening and climate is that by Franklin[18] which covers many applications of meteorological knowledge, mostly from the author's wide experience.

12.8 Human Health

Climate affects the health of man in many ways and throughout his life, and for a full description of the subject of medical bioclimatology the reader is referred to a monumental work by Tromp[19] who has given an excellent summary of knowledge of this science. A more general text, edited by Licht,[20] offers a good introduction, through its twenty-eight chapters, to the field of medical climatology.

The effects of climate are, according to Mills,[21] playing their part even at the time of conception, for he states that 'babies conceived during the very hot months lack the vitality of those conceived in the cold months'. Clearly the state of health of the mother also plays a fundamental role in the baby's health at birth and, once the baby is born into the world, climate begins to exert an influence which may be physiological or psychological and direct or indirect. In this section we shall discuss only the physiological aspects since there is, at present, much controversy and uncertainty concerning the psychological influences, some of which are of a conjectural nature.

12.8.1 *Direct effects* (detrimental)

Radiation. If there is an appreciable increase in the amount of ultraviolet light normally received by the body, such as occurs on high mountains, intense sunburn, conjunctivitis and cataracts can result. It is thought, also, that a mild form of skin cancer is caused in white-skinned people living at high altitudes in the tropics. The white settler in these areas also tends to perspire a great deal and skin ailments become difficult to cure and heal. Intense radiation can lead to heat stroke whether or not there is an increase in ultraviolet radiation.

Temperature. People living in the extremely hot regions develop remarkable tolerance to the searing heat of the soil surface. People in Arab countries, for example, can walk barefooted on sand at a temperature of 70 °C, a temperature at which the sandal-clad European begins to feel uncomfortable (a fact to which the author can attest). Similarly, in intensely cold regions, the inhabitants can walk barefooted in the snow all day with no sign of discomfort. The Andean Indians have astounding resistance to cold, their feet being well endowed with small capillary blood vessels through which heat is circulated to the feet rapidly. An investigation by LeBlanc[22] showed that the Gaspé fishermen of Canada have a general peripheral adaptation to cold.

In conditions of extreme cold, frost bite of the extremities is common, but the worst effects are usually on the lungs of the inhabitants. Pneumonia and other pulmonary diseases are common in the highlands of Bolivia where the cold season is known as 'the harvest of death'.

Temperature and humidity. High temperatures combined with high humidity can lead to outbreaks of 'prickly heat', an intensely irritating rash that, once established, can be cured only by moving the sufferer for a short time to a healthier environment. Heat stroke can result if conditions are severe, and Schikele[23] has shown a pronounced increase of heat stroke when the temperature (°C) exceeds $(48 - 1 \cdot 1e)$, where e is the water vapour pressure in mmHg.

Marine climates with high humidity and raw winters appear especially conducive to such complaints as rheumatism and arthritis. Intense cold combined with moisture can increase pulmonary diseases while the combination of cold and saturation of the clothing can lead to unpleasant complaints, such as trench foot, when the shoes get saturated, or to death, if much of the clothing is soaked.

Warm or hot conditions with very low humidity, such as can occur in the tropical highlands, lead to a chapped skin, the cracking of lips and pronounced bleeding from the nose.

Pressure. It is considered that it is impossible for people to live continuously at elevations above 5 200 m. Experiments in the Andes have convinced mine owners that the workers have to be transported daily to work the mines at 5 800 m since the rarefied atmosphere leads to an oxygen starvation that cannot be corrected by acclimatization. The highest town is Wenchuan, at 5 100 m, and the inhabitants of the Chacaltaya High Altitude Laboratory in Bolivia, at 5 250 m, remain there for only about one week at a time.

Smog. Smog is usually considered a meteorological phenomenon and so it is mentioned here as a direct cause of ailments. More correctly, however, it should be listed under indirect effects since the source of the trouble is actually the introduction of foreign bodies into the system. Smog can be one of the deadliest of meteorological phenomena because it concentrates the foreign bodies in the air which people breathe. Thousands of people, mainly the elderly and very young, died from pulmonary diseases during the London smogs of 1952 and 1962.

Intense disturbances. Since we are concerned here with health it must be remembered that such phenomena as tornadoes, cyclones, hurricanes, typhoons and floods associated with torrential rains can lead to the loss of many thousands of lives. The worst flood ever recorded occurred in China in 1887 when nearly one million people perished, and the worst typhoon disaster took 300 000 lives in Indo-China in 1881. Under this heading we may include some of the famines that have been attributable to the weather, such as the blight (10.3) which destroyed the Irish potato crop in the 1840s.

12.8.2 *Direct effects* (beneficial)

It is fortunate that not all the effects of climate are detrimental, but it is not easy to list the ways in which climate is of direct benefit to health. It is realized that some climates are more stimulating than others, tending to toughen the individual by the rigours of the weather to which he is subjected, while others are, unfortunately, enervating, and sap the strength of the populace.

Radiation or sunlight is perhaps the most important beneficial element of climate since ultra-violet radiation devitalizes some bacteria and germs; it also is antiricketsial. Sunlight can also be a psychological benefit, especially after long spells of overcast conditions.

12.8.3 *Indirect effects*

Climate and weather have a great effect upon many of the carriers of disease and we may consider these under two subdivisions, air- or insect-borne pathogens.

Air-borne pathogens. Allergies are rather an unknown quantity. Nearly everyone appears to be allergic to some thing or condition and, although they have not been fully investigated, many of the identifiable allergens, such as pollen, spores, etc., are known to become more concentrated or to travel farther under special climatic conditions.

Insect-borne pathogens. Malaria, yellow fever, dengue, filariasis (elephantiasis) are all communicated by mosquitoes whose breeding, life cycle and movements are dictated partly or completely by climatic conditions.

Many insects or arachnids (mites and ticks) are carriers of disease such as bubonic plague, tick fever, sleeping sickness and innumerable others. Movements of these animals are often dependent upon climatic conditions. Some diseases are only endemic to certain regions because of the suitability of the climate, for instance bilharzia.

Diet. Another indirect effect of climate upon human health is the role that it plays through the diet, causing a weakening or strengthening of the human organism to disease attack. It is particularly noticeable in the case of babies, old people and convalescents.

12.8.4 *Miscellaneous effects*

There are some effects which are believed to be correlated with certain atmospheric changes of small amplitudes. Such relationships include apoplexy, and acute closures by gall and kidney stones which are twice as common on days with pronounced air mass changes as would be expected by chance; suicides are greater on days of passage of fronts; intermittent exposure to negatively-ionized air appears to give some sedation and relief of pain on the first and second post-operative days. There is also suspected to exist a relationship between crime outbreaks and weather, and Brezowsky[24] has shown that the foehn wind causes an increase in industrial accidents.

Chapter 13

CLIMATE AND ANIMALS

13.1 Introduction

Man has always thought that certain animals were possessed of special knowledge, not granted to him, a type of 'sixth sense' that enabled them to foretell certain elements of the weather. Literature and folk-tales are full of small items of weather lore that depend upon the reactions of all types of animals, and these tales are prevalent in all parts of the world, civilized and uncivilized. It is strange that so very few of these ancient beliefs have ever been tested in order to ascertain whether or not they are 'old wives' tales'. It is likely that the beliefs entered into literature because of mankind's inherent wish to believe in something that transcends the law of nature as he (man) understands them. It is easy enough to remember the times when there appears to be a connection between certain behaviour and the subsequent weather conditions and to forget the occasions when there is no correlation. One of the early meteorologists, Dr C. C. Abbott, did actually observe two aspects of animal behaviour, the building of retreats by muskrats and the storing of food by squirrels, but he found no correlation with the weather following their actions. On the short-term approach, though, the answer is different. It is known that some animals do have what is called a 'sixth sense' in that they are very much more sensitive than humans to small changes in pressure, temperature and humidity. This enables them to respond to these changes by their movement so that they are then giving a very short-term forecast: one that man can use, if he is not already possessed of the information by his own efforts via instrumentation and scientific knowledge.

The direct influence of climate on animals is usually small. It normally has an indirect influence through the plant community, the food source for most animals, even the carnivores being dependent upon the existence of the herbivores within an area to supply them with their normal diet.

In spite of this it is possible to find some correlation between types of animals and climatic areas but, if one looks a little further, it is again effected through the plant community. For example, in tropical rain forests, which have three- and four-tiered growth, the main sustenance is available in the upper layers. The large animals which live in them must therefore be able to climb or fly.

In the temperate grasslands conditions are very different; there is abundant herbage during the summer combined with a severe winter and lack of food. Few animals can tolerate these changes and the denizens of these regions are either birds and large animals that migrate in the winter or insects and burrowing animals that retire to a less severe winter habitat.

For the purposes of this chapter it is convenient to consider the animals as subdivided into five groups, the mammals, flying animals (birds and insects), aquatic species, reptiles and those animals to be considered as being within more than one subdivision depending upon developmental stage or the time of year.

A consideration that has been only touched upon in general investigations is the response of animal vision to various wavebands of electromagnetic energy; in other words, comparison of animal sight with humans. It is known that most insects can see much farther into the infrared than humans but are blind in the blue region.

13.2 Mammals

Mammals are warm blooded animals (homeotherms), that is, they all possess a thermo-regulation that strives to keep their body temperature within a narrow range. For the larger mammals (man, cattle, pachyderms) it appears that this temperature is around 38 °C, the tolerance limits being small so that a few degrees' rise or fall will often prove fatal to the mammal. The critical upper threshold ranges from about 39 to 45 °C

Comparison with man enables us to state that mammals gain heat from four main sources: metabolism, radiation, convection and conduction (Chapter 12).

They lose heat by four channels: radiation, convection, conduction and evaporation.

It must be noted that all mammals, save the rodents and lagomorphs, have sweat glands, although it is not always known whether or not these function in the same way as for humans. In some cases it is suggested that they do not function at all!

13.2.1 *Mammals in a natural environment*

Animals find life a constant struggle against the forces of nature, forces that include climate. Therefore under the natural conditions that we are presupposing the animal has usually come to an equilibrium with his environment and is able to survive within the specific climatic region. Of course, owing to the vagaries of climate and weather, extreme conditions sometimes occur and either directly or, more usually, indirectly, by the effect on food and water supply, contribute to the destruction of the weaker members of the species. If the conditions are really severe, such as can occur in, say, the tropical savanna regions when a long period of drought occurs, then vast numbers are threatened with extinction. Many of these animals follow fixed migration routes on an annual cycle and the failure of rains in some area can lead to the death of thousands of beasts.

Few data are available concerning physiological and climatological aspects of life for the animals in natural habitats and often observations of behaviour are explained, or not explained, by relating them to weather conditions that are not representative of those in which the animals were living at that time. The problem is very complex and depends upon a real understanding of microclimatic variations and their patterns. A few of the results are given here.

The pachyderms, elephant, rhinoceros and hippopotamus, have body temperatures around 35·5 °C, the rhinoceros showing a marked diurnal variation of about 3·5 °C. The hippopotamus sweats very profusely whether the weather is cool or hot and whether the animal is resting or moving.[1] For this reason it is necessary for the hippo to spend much time in the water or mud so as to effect the cooling of the body by using an external water source and reducing the radiation heat load. The elephant, although not being subjected to such a stress, also enjoys the occasions when it is able to douse with water and experience the rapid cooling effect.

In studies of the camel[2] it was noted that the diurnal variation of the rectal temperature was normally about 2 °C, but when deprived of water it rose to 6 °C. This tolerance is most important for it means that the heat can be stored and water does not have to be evaporated to produce cooling. The fur appeared also to play a useful role since a shorn animal lost more water by sweating.

Mammals normally have fewer pathological enemies in the cold climates than in the tropics but the reader is advised to consult specialized publications if wishing to pursue this aspect.

At the other extreme of size the white rat suffers very much at temperatures of 30 to 35 °C with a humidity of 30 mmHg. One experiment noted that 30 per cent of the animals died, while at the same temperature with a humidity of only 10 mmHg they were not under any apparent stress.

A recent paper by Johansen[3] gives a review of the knowledge of the responses to heat and cold in the lower mammals. He notes how the monotremes (echidna and platypus) can vary their heat production, but for heat loss, when the air temperature is above that of the body, around 32 °C, they must depend upon cooling by external water sources or burrowing. They possess a very small number of effective sweat glands. The marsupials, with a body temperature of about 36 °C, can vary both heat production and loss. They possess primitive sweat glands but are noted to lick parts of the body to induce evaporation and cooling at temperatures above 38 °C. The insectivores have a very labile body temperature with pronounced diurnal variation. Within the group the body temperature varies appreciably; for instance, in the shrew it is 41 °C and in hedgehogs from 31 to 36 °C. The chiroptera or bats form an interesting group; the microchiroptera of the cold regions alternate between a homeothermic and a poikilothermic (cold-blooded) state as they go from daylight inactivity to night-time activity. When the body temperature is regulated it is higher than for the marsupials. When in the dormant stage most bats seek out a relatively warm equable environment.

Some animals display a change of coat with climatic variation. Dwellers in the cold regions often develop a white coat during the snows of winter, while others, such as the Himalayan leopard, simply grow a longer fur during the cold season.

13.2.2 Domestic animals

Domestic animals must be considered separately from mammals living under natural conditions because:

(i) breeding and domestication have usually made these animals unable to withstand the normal rigours of the climate in which they dwell (beef cattle are an exception to this);

(ii) their food supply and shelter are generally guaranteed by their owners;

(iii) owing to their prime importance to either the food supply of man (cattle, sheep, pigs, goats, etc.) or their service to man (dogs, cats) they have been the subjects of more intensive study.

Animals that are kept because of their food production have a yield dependent upon climate, management, nutrition and breeding. Management and breeding considerations are outside the scope of this work but nutrition is clearly bound up with the climatic aspects. Production is less in the hot regions of the world in all the factors of meat, milk, skins, wool, hair and power. The climate can of course be acting through parasitism, which is generally of much higher incidence in the hot, humid regions. In these same regions at certain times of the year the grass looks deceptively lush, but in fact it is poor in protein and high in fibre content. In spite of these indirect effects a good percentage of the diminution in productivity must be attributed to the direct effect of the climate upon the animals.

As we are considering the effect of heat stress and not the effect of extreme cold conditions, we are concerned therefore with the heat tolerance of animals. This heat tolerance is not an easy thing to measure or even assess with any accuracy for it is an integration of so many factors. In Chapter 12 we noted how difficult it was to prepare a single reliable comfort chart for man, an animal who can say what his feelings are, imagine then the complexity of the task when dealing with other animals.

First it is necessary to define heat tolerance and many workers have assumed this to be simply the measurement of the deep body (rectal) temperature of the animal, increases of this being directly related to the heat stress it is undergoing. This is perhaps a good first approximation but does not tell the full story. Other workers have used respiration rate. When this increases the animal is attempting to lose heat by evaporation of water from the respiratory organs and tract. Yet others have used the concept of respiration volume. It is then possible to draw graphs of temperature and humidity combinations v. rectal temperature, respiratory rate or volume for any specific breed or type of animal; but is this a true estimate of their heat stress or their comfort? These experiments are carried out mostly in a special climatic chamber, with or without radiation load possibilities. If it is without radiation it is surely not 'natural'; if it is, then the radiation is not of the spectral distribution of sunshine, which must have a different effect.

Some investigators have tried to pronounce upon the suitability of certain climates for the rearing and economic production of certain breeds of cattle by use of the climagram or the hythegram. These are simply diagrams that show the month-by-month variation of temperature and humidity or temperature and rainfall respectively. Bonsma,[4] for example, has proposed that if six or more of the monthly points fall within a zone drawn on the graph (he uses four zones altogether), then cattle corresponding to that zone may be reared there satisfactorily. The method is seen to have its limitations but gives a rough indication for practical purposes.

Cattle. Cattle have been the subject of more study from the point of view of their response to climate and weather than any animal, save man, and it is impossible to do more than summarize present thought upon the subject.[5,6]

Beef cattle are normally tough animals and the

major effects on them are through the pastures and their water needs. Dairy cattle are more limited and delicate; they are also generally housed for those parts of the year when they might experience a major climatic stress.

Cattle gain heat from:

metabolism — a 20-month old cow generates 2 400 kcal/m^2 day which rises to 3 800 at the height of lactation. If a cow walks 1 km, say to water, she generates an extra 45 kcal/100 kg weight;

radiation — this load depends fundamentally upon many factors but in the hot dry areas it is likely to be about 75 + 11 (T − 35) in kcal/m^2 hour, where T is the air temperature (°C);

convection and conduction — not a great deal is known about these aspects.

Radiation loads can be appreciably reduced by the supplying of shades, both artificial and natural. Such shades must not prevent the flow of air over the animal. For artificial shades (white top with black underneath) it has been shown that, with high radiation loads of about 2 cal/cm^2 min, if the shade is at a height of 3 m, then at 60 cm from the roof the radiation load is reduced to 36 per cent of the exposed value, increasing to 38 per cent at 2 m and 45 per cent at 2·3 m. Clearly, the existence of some shade will greatly add to the 'comfort' of the animal when there are times of heat stress.

The general problems of housing for cattle are numerous. If iron roofs are used then condensation and drip-off may occur leading to a resultant reduction in the rate of fattening of the animals owing to the presence of water on the skin. Beef cattle can exist with only crude shelters and wind breaks even down to temperatures of −40 °C. It is interesting to note that at high temperatures the sky is normally cooler than the air and therefore if the animal can be exposed to it there will be a greater heat loss by radiation. For this reason the surface area of the shelter should be just sufficient to protect the animals.

The other main source of heat loss is through evaporation. To help the animal various methods have been advocated such as sprinkling or supplying cooled water. The animal, in order to lose more heat by evaporation, normally begins to pant quickly under heat stress conditions, increasing from its normal rate of 20—24 times per minute to

about 150—200 at 40 °C. Tropical cattle lose a lot of water by diffusion through the skin or sweating. There is still much discussion as to whether or not the sweating process is effective in losing much heat for the animal. Brody[7] gives the following table.

Percentage of heat production dissipated by vaporization

T (°C)	European breeds	Indian breeds	Man
10	23	29	18
21	50	36	29
32	79	78	66
41	104	112	194

The hair of cattle acts rather as clothing does on man, in that it tends to reduce the radiation load and also to trap air and thereby insulate the animal. The degree to which it affects the balance depends upon its colour and thickness.

With respect to management some aspects of climate such as precipitation, wind and sunlight, can be important in helping to maintain sanitary conditions in pastures or feed-lots. Also, in the same way as man can acclimatize, it is possible to breed cattle for tolerance to certain climatic conditions. It should be noted that high environmental temperatures can cause a decrease in fertility.

The pests of cattle are themselves affected by the climate, and especially is this true of the parasites; for example, the internal ones such as worms must pass the winter in their host and can then be treated. External parasites, lice and ticks, find it hard to survive extreme heat in summer and are then at a minimum.

Sheep. Sheep have not been studied quite as much as cattle and less is known about their heat balance and water balance problems.

A basic difference between cattle and many breeds of sheep is the fleece of the latter. Because of this the sheep can sustain themselves even in very low temperatures (−45 °C) with the protection of only minimal shelters. The fleece acts as a very effective insulation, trapping a thick layer of air near to the animal's skin. Owing to this insulation sheep can survive when buried beneath snow but if snow packs into the fleece when the herd crowds together some animals can be suffocated. High winds tend to flatten the fleece and reduce this insulating layer. The long fleece of the merino breed also aids in heat dissipation by giving an extra protection against solar radiation; however, short-coated sheep can also thrive under the same conditions.

The rectal temperatures of sheep are about 38 °C but at temperatures above 29 °C with a r.h. of over 65 per cent the rectal temperature and respiration rate rise very appreciably.

Some breeds of sheep such as the black-faced Persian, living in the semi-arid areas, have a very large tail that acts as a store of fat during the periods of drought and poor grazing. At the end of the drought the tail is just skin and bone.

Newly-born lambs are very susceptible to cold and snow and must be protected against extreme conditions.

At the time of shearing the fleece must be dry, otherwise the wool cannot be bagged and the shearers get 'cramps'. The newly-shorn sheep, if exposed to rain or snow during the first day or two, can develop a type of pneumonia.

Pigs. Pigs are the least tolerant of the productive domestic animals, a fact that has been appreciated even in many undeveloped areas where mud shelters have been satisfactorily built for their protection. Pigs have a rectal temperature of around 38 °C, but show a much greater variation (35–40 °C) than sheep and cattle. Their comfort is improved if water for wallowing is available. They have sweat glands similar to those of cattle.

13.3 Flying Animals

Under this heading we shall consider those animals, such as birds and insects, that, at some time during their life cycle, have the ability to fly. The insects are cold blooded and show a body temperature dependent upon environmental conditions. When the animal is in the stage at which gliding or flight is possible, it is subject to the effect of wind currents and each animal will have a threshold wind speed at which it finds itself unable to make headway against the wind pressure. In general, this threshold pressure will be a function of the shape of the animal in flight, its surface friction with the air and the power of the animal. For instance, locusts find it almost impossible to hold to a course if the wind speed exceeds 16–20 km/hour.

13.3.1 *Birds*

Birds do not have sweat glands and, short of wallowing in water, birds exposed to extremely high temperatures and intense radiation have either to ascend into the air to find the cooler upper layers, to pant excessively or to find water in which they can bathe and benefit by evaporative cooling. Normal body temperature ranges from 41 to 44 °C while the highest body temperature for survival is about 46 °C. Their feathers act as a very effective insulation for they can puff themselves up to a large relative size and hold a layer of air close to themselves. This way they can withstand low temperatures but when freezing conditions exist birds have a very hard time in finding food and water for their small needs and human influence can be of great importance at such a time.

Bird migration is one of the great unsolved problems of nature and this book is not going to attempt to explain any of its features in terms of climate and weather but it must be noted that extreme winds and storms can cause whole 'squadrons' to become exhausted and die at sea or in the deserts and other inhospitable areas.

The domestic fowl, owing to its economic value, poses a special problem and it is desirable to provide it with certain forms of shelter. Lately, the battery- or cage-reared bird which does not experience a natural environment has become popular. Light plays a very important part in egg production since its penetration of the eye sets up a nervous stimulation; it is in fact one of the triggering mechanisms in egg-laying. The length of the natural daylight is also of importance since increasing day length will bring the pullet into production earlier than it will for decreasing day length fowls. Under artificial conditions this aspect can be controlled.

The domestic fowl has a body temperature of about 42 °C, with an appreciable diurnal variation. At high temperatures of about 32–35 °C, either outside or within the battery, heat prostration can result.[9] Ideally the temperature of the fowl's environment should not exceed about 27 °C, and to ensure this shading, cool water and adequate ventilation should be supplied. It is also desirable to provide night cooling, as with low summer minima, say about 15–18 °C, the fowls do not seem to suffer as much discomfort. When heat stress conditions arise the birds usually extend their wings so that more of the peripheral flow of blood can be cooled.

Temperatures above about 21 °C appear to have an effect upon egg size and where this temperature is exceeded for a period of days there occurs a decline in the egg size. High humidity allied with high temperatures brings about a reduction in shell thickness.

13.3.2 *Insects*

Being cold blooded and having a body temperature dependent upon the environment, insects are usually relatively unaffected by extreme weather conditions. There is a most interesting paper by Parry[10] in which the author states that the temperature of arthropods in sunlight is determined by the balance between radiation and convection heat exchange since metabolism and evaporation are normally unimportant. Because the convective loss varies approximately as the 0·25 power of a linear dimension, the smaller the animal the lower will be the temperature at which the convective loss will equal the heat load. When the insect becomes unable to hold a balance the body will heat and temperatures high enough to damage the chitinous wax and other delicate parts can be reached. Such conditions can occur to young locusts, hoppers, under extreme heat conditions in the desert when the radiation load is great, while harvester ants cannot survive a ground temperature of 50 °C.

Insects that move only short distances often use the ability to fly in order to pick out the most suitable microclimate in their environment and Geiger[11] cited the example of the insect horde that ascends to the warm layer below the night inversion when the sun's rays first begin to penetrate the forest or wooded areas. Similarly, the mosquito can cruise around to find the pools of stagnant water that it needs for laying its eggs, pools that owe their existence generally to heavy rainfall.

Animals that can fly long distances, such as adult locusts, are dependent upon the vagaries of the weather and the important work of Rainey[12] has shown how the desert locust (*Schistocerca gregaria*) is drawn into areas of converging winds. Such meteorological conditions are associated

with rainfall so there is a pronounced correlation of locust incidence and rainfall of this type.

Kleem[13] has identified six layers in field work, each of which has a different microclimate and a different fauna:

geobium – beetle larvae, butterfly pupae
herpetobium – beetles, spiders, ants
bryobium – mites and spring tails
phyllobium – arthopters, aphids, caterpillars
anthobium – flower visitors
aerobium – libellae.

Insects are very responsive to the meteorological environment so that an increased appreciation and knowledge of its effects upon them must be realized if we are to understand, extend or refine their control. Our knowledge must extend not only to the direct effects of all the weather factors but also to the indirect effects acting through the hosts, parasites and predators.

There are many climatic parameters that have been shown to be correlated with certain insect incidence or behaviour. The following list is but a small sample:

(*a*) the tree cricket has a rate of chirping related to the air temperature, T (°C),

$$T = 4·5 + 0·14 \text{ (number of chirps per minute)};$$

(*b*) heavy and intense summer rains destroy chinch bug nymphs;

(*c*) a prolonged spell of high temperatures leads to an increase in the number of house flies;

(*d*) a delayed spring is favourable to the corn-seed maggot while a hot, dry spell terminates an outbreak.

(*e*) The first rains after a long dry spell seem to presage the advent of the flying stage of termites.

Insects can tolerate a great range of conditions; for example, some mosquito larvae, water bugs and water beetles can exist normally in water at 38–50 °C, while the pioneering work of Glick[14] has shown the incidence of insects in the free air at heights of up to 4 200 m – many of these were wingless insects such as springtails (3 300 m) and fleas (600 m).

13.4 Aquatic Animals

Animals dwelling in water experience a much more stable environment, from the climatic viewpoint, than those inhabiting the land. The only climatic element thought to be of importance to the aquatic breed is temperature. The annual range of temperature in seas is seldom above about 14 °C, the extreme values ranging from about 30 °C in the Arabian Sea to −2 °C in the polar seas. Because of the very large heat capacity of water, temperature changes take place slowly and most aquatic animals, being conditioned to this, will die if subject to rapid changes in temperature. The marine animals tend to accept only small changes of temperature and during the summer months will exhibit movements either polewards or downwards, seeking out cooler waters. During the winter many species inhabit the lower layers where the temperature is often some 3–6 °C higher than the surface layers.

A special effect of climatic factors upon fish is the alteration of certain ocean currents. When this occurs the plankton on which the fish live die rapidly, causing decreases in both the fish and the fish-eating bird populations.

Animals living in the shallow waters or along the seashore, experience slightly more variable temperatures, while rock-pool life has sometimes to survive quite high temperatures. However, since the changes come gradually all marine life finds them tolerable. Crabs that live in shallow holes will move into pools if the sand becomes too hot.

13.5 Reptiles

These cold-blooded animals, comprising crocodilians, lizards, snakes and turtles, possess scales that help to reduce the evaporation of their body water. They maintain blood temperatures within an optimum range, a range greater than that of the homeotherms, by seeking out the environment necessary either to raise or to lower this temperature. During the heat of the day these animals often burrow underground or seek shady spots. It appears that the optimum blood temperature is about 36–38 °C and many lizards suffer heat prostration at 40–47 °C; death resulting if the blood heats above about 50 °C. For many desert snakes the critical level is lower by about 5–8 °C. When it is appreciated that the soil surface temperature in the sub-tropical desert habitats of many of these animals can reach 80 °C, it is clear that the animal must seek other environments of a more hospitable nature. The basking of lizards is an attempt to

maintain optimum body temperature. Even the tough-skinned crocodilians cannot stand too much heat and must enter the water to cool − if denied access to shade and water the crocodilian will die.

An interesting response to heat and light is demonstrated by some chameleons, the body turning a darker colour when exposed to the sun while shaded portions remain light-coloured.

13.6 Soil Dwellers

Within the soil many animals find the equable conditions they need to escape the extremes of the open environment. Some animals, such as worms, spend their whole life below ground but most only use the earth as a supplier of acceptable environment.

During the heat of summer some animals, ranging from insects to mammals such as the desert fox, burrow underground to escape the intense radiation load during the day. Pruitt[15] investigated the Alaskan taiga during winter and spring, and has provided measurements which illustrate the less extreme temperature of the subnivean environment. For instance, the absolute minimum temperature at 8 cm depth below the mossy floor was −15 °C, while at 22 cm it was −5 °C.

In winter, in the cold zones, there is a movement underground during the season of hiberna-

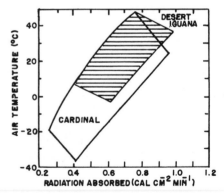

Fig. 13.1. Climate space diagram for a cardinal and a desert iguana showing the combined limits of air temperature and of radiation absorbed by the surface of the animal within which the animal must live when the wind speed is 100 cm s⁻¹. Increased amounts of absorbed radiation require that the animal is restricted to a lower temperature. (After Gates[17].)

tion. Even such large mammals as bears take part in this phenomenon, finding below the surface the constant conditions so necessary to the hibernating animal. Scholander *et al.*[16] have studied the best regulation in some active animals and note the insulation value of the burrows and fur against the intense cold conditions.

It has been expressed very well by Gates that 'an animal is coupled to the climate surrounding it by energy flow.'[17] Any animal, including man, must remain in energy balance with its environment most of the time. This means that knowing the metabolism, water rate loss, size, absorptivity and emissivity to radiation, insulation and body temperature requirements of the animal, plus the limits of tolerance of these quantities then it is possible to predict the combinations of radiation, air temperature and wind speed within which 'climatic space' the animal would be in thermodynamic equilibrium. An illustration is given in Fig. 13.1, with an assumed wind speed of 1 m/s, and it is seen that the cardinal can survive at far lower temperatures than the iguana but will have to seek shade at temperatures above about 30 °C, while the reptile can tolerate high radiation at that temperature.

Chapter 14

CLIMATE AND BUILDING

14.1 Introduction

The first men on earth, the forerunners of *Homo sapiens,* lived naked and exposed to the elements. Gradually, however, as they moved into other areas, as the natural hair covering of their bodies decreased and they became 'softer', they felt the need to protect themselves from the vagaries and rigours of the weather. Clothing is dealt with in an earlier chapter, but here we deal with the equally important aspect of man's attempt to protect himself, the obtaining of shelter.

When technical knowledge was almost non-existent, man was not possessed of the ability to build habitations and had to be content with the shelter supplied by nature, usually in the form of caves. He soon recognized that within these caves he was protected, not only from an omni-directional attack by enemies or wild animals but also from the weather, however inclement, so that he was still relatively safe and snug within his 'home'. We know now that the climate within caves is very equable, most climatic variations being smoothed out within 15 m of the entrance. At first man would have had to exist in the cave mouth region where he could get sufficient light during the sun hours. Later as he was able to control fire and heat, he could retreat farther to the protection of the cave to enjoy the uniform climate there. In Egypt, Sutton[1] demonstrated how the diurnal range of temperature varied from 40 °C outside, to 11 °C in the cave entrance, reducing to less than 3 °C at 150 m inside the cave. In the hot regions where man is first thought to have have existed the very warm external temperature would have been exchanged for a low temperature with a 100 per cent relative humidity, according to Buxton[2] and Polli.[3]

If we take the Australian aborigine as the next development example, we see that man then began to build himself crude shelters, more in the nature of windbreaks that were not vertically erected but were angled so as to reduce the night radiation loss. Tents soon came into being when the nomadic pastoral peoples needed to be often on the move, following the rain with their herds. These covers gave protection from the precipitation and radiation while allowing through ventilation when this was required for cooling.

When man became more settled he turned to the naturally available materials of his region, either wood or stone, to build himself and his family a protection suitable for all seasons. Over the years he has gradually developed the optimum design, within his limited means, with regard to technology, materials and cost, to combat or minimize the extremes of his region's weather. We can still learn a lot by taking note of some of the design characteristics of the indigenous people.

While man has been able for many millennia to supply artificial means of heating in order to combat the cold spells, it has only been within the past few decades that he could supply cold air and reduce the stresses due to intense heat, either temperatures or intense radiation. It must be realized though that the modern method of 'air conditioning' is limited in application because of its cost. The approximate cost of cooling an average-sized house to a suitable level (say about 24 °C) when the external temperature is high (35 °C) is around $1 (50p) per day, in a region where electricity is relatively cheap (USA) while the cost of the apparatus is about $400 (£200) per living room. Although it is in the USA where air conditioning has achieved its greatest hold, the only completely air-conditioned community is in Kuwait. Because of its high cost and therefore its economic limitations, it is necessary to make full use of all sensible, cheap methods of achieving cool comfort, such as making the most of cooling breezes, cutting out intense radiation loads, installing efficient insulation, and protection from hot winds.

The importance of the effect of the weather on buildings is realized nowadays by most architects. Professor Page of Sheffield University has been a leader in this field and his review[4] of the problems is a good starting point for interested readers. Realizing the fundamental nature of the problems UNESCO supported a special symposium during the Third International Biometeorological Congress in 1963 on 'Indoor Climate in Arid and Humid Zones'.[5] More recently the UNESCO have published a text concerned with climate and house design that is especially useful for warm climates.[6]

14.2 Climate and the Architect

When an architect is faced with the problem of preparing the plans for a specific construction he is made aware, at a very early stage, of such fundamental factors as the type of construction required, house, church, offices, shop, etc., the location of the building and the cost limitation. It is most unfortunate that he is not also supplied with equivalent climatological knowledge. Too often the appearance of the finished product is put ahead of its utility, while shortcomings of the design are excused on the grounds of aesthetics. Now that the designer is faced with scientific aspects of construction, it is time he accepted that some climatological factors must be appreciated and allowed for if he is to provide well-designed buildings at economical rates. This does not, of course, mean that the aesthetic aspect must be lost for this is where the true 'art' of the architect shows through, but this is a sphere where art and science must be blended.

Atmospheric conditions are very relevant to a number of aspects of construction, choice of site, optimum climatic data needed by the architect may be considered under four main headings.[8]

(1) thermal considerations
(2) ventilation and wind pressure
(3) daylighting factors
(4) precipitation or dampness aspects.

14.2.1 *Thermal considerations*

In order to judge correctly the thermal load upon a building it is necessary to know the relative magnitudes of the elements in the heat balance equations, and to know these for the different seasons of the year. Thus, ideally, the architect should know air temperatures, solar radiation, humidity and air speed, plus the frequency analysis and interrelationships of these variables. Unfortunately, it is only in small areas of the world that this information is available, and then not necessarily within the particular microclimate of the site. Thus, the ideal situation is rarely realized in practice, usually only on specific research projects. Siple[9] has presented an excellent graphic representation of the climate of certain cities that should be in more general use by architects.

In practice, the architect will have to content himself with a knowledge of the broad climatic features of the region plus a common-sense approach to a realization of the variations introduced by microclimatic factors at the particular site.

It has been pointed out in earlier chapters how the microclimate of an area depends upon the aspect, slope, vegetation cover, etc., and it is not to be expected that the architect be aware of all the subtle but very important climatic changes that may occur over a short horizontal distance. Consultation with a competent climatologist will soon put him in possession of the relevant details concerning both the regional or macroclimate and the local or microclimate of the site.

Knowledge of the thermal characteristics, plus on a secondary level the humidity, is the most important detail, and then the problem arises – just how should the data be presented to the architect?

Perhaps the best method is to have the temperature plotted against time so that, at a glance, it is possible to read just how much time the temperature is above, or below, a certain threshold (Fig. 14.1).

Generally, such is not the way that the temperature data are presented but useful estimates can be made from either the mean monthly maximum and minimum temperatures or from the mean of the highest and lowest each month.

14.2.2 *Ventilation and wind presure*

Wind roses of speed and direction for each month (Fig. 14.2) are required if accurate information is to

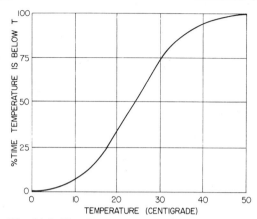

Fig. 14.1 The distribution of temperature with time

be given about ventilation and wind presure on a site. Information from a near-by station is not directly relevant, but it will enable the climatologist to supply data concerning the prevailing winds, wind pressures, etc. It will only be an 'on the site' inspection that can decide such things as: is the building situated where it is sheltered by natural or man-made objects, or is it excessively exposed?

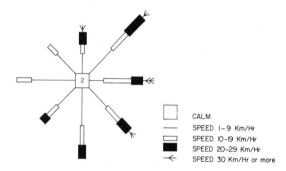

	CALM
—	SPEED 1–9 Km/Hr
▭	SPEED 10–19 Km/Hr
▬	SPEED 20–29 Km/Hr
←	SPEED 30 Km/Hr or more

Fig. 14.2 A theoretical wind rose

The normal increase of wind speed with height must be appreciated by the designer. It must be realized that the theory usually applies only to regions in which the wind has a long fetch over fairly uniform surfaces; it does not apply in towns or densely built-up suburbs. Thomas and Dick[10] have suggested that for the first few floors in a central town exposure, the wind speed is but one-third of that in the free air, increasing to about two-thirds for the middle floors and suburb locations, and unity for the upper floors. It may be that for site

investigations the best indication of the wind pattern in a built-up area may be given by a study of the air flow through forest stands and clearings.

The pressure of the wind upon buildings is another very important consideration, for the building should be designed to withstand the maximum likely winds. The pressure of the wind is proportional to the square of the wind speed multiplied by a factor depending upon the shape of the construction. For a house the dynamic wind pressure is about 14 g/m² for a wind of 1 km/hour, rising to about 25 kg/m² in a 50 km/hour wind. Evans[11] has given details of air flow patterns around houses of various shapes.

Work by Thom[12] has shown that information concerning the annual extremes of the 'fastest mile' observations provides the best available series of data for design purposes. Thom also presents a formula for reducing these series to any chosen height above ground.

14.2.3 *Daylighting factors*

Under this heading we are considering the problems of natural lighting patterns within buildings and it would appear, at first, that climatic values would play an important role. There are very few values of daylight measurements in the literature, and information would have to be obtained by using a conversion from radiation or even sunlight data.

The normal approach to this problem is to use a nomogram or template of the solar altitude and azimuth so as to calculate the sun's position and then to compute various daylight factors within buildings by means of graphs or further nomograms. Certain assumptions, based on work by Moon and Spencer[13] and by Petherbridge,[14] concerning the relative brightness of different sectors of the sky are used in these calculations.

Practical results that obtain from these calculations include such matters as the siting of clerestory windows and the location and size of an overhang for cutting out the high sun's rays while permitting the heat from the low altitude sun to enter the room.

14.2.4 *Precipitation or dampness aspects*

Rainfall has an appreciable effect upon the exterior of buildings, affecting the materials, water-

proofing, damp courses, etc. When it is coupled with
with high winds it may also affect the interiors. At
present, little is known about the incident angle of
the falling rain, but an investigation by Lacy[15]
showed that, in the London region, this varied
from 30° to the vertical in winter to 15° in the
summer. In many tropical downpours the rain falls
very nearly vertically. By making certain assumptions about prevailing wind directions it is possible
to construct overhangs and protective walls in
order to shelter the exposed house walls from the
brunt of the rainfall.

Thein[16] has studied the penetration of rain into
houses and has derived the formula:

$$p = rv^2$$

where r = maximum rainfall in 5 minutes, mm,
$\quad\quad v$ = wind speed in m/s at the time r occurred,

stating that rainfall penetration spots will occur if
p exceeds 100.

This equation stresses the importance of rainfall
intensity over a short period as a fundamental factor.
The designer should allow for adequate roof and
ground drainage since a very large volume of water
can soon gather from a large collecting surface. For
example, 5 mm of rain over a 50 m^2 roof amounts
to about 0·25 m^3 of water (225 kg) and would fill
8 m of 10 cm wide guttering to a depth of 30 cm.

Knowledge of the flood plain levels must also
be considered, for during the life time of the building some rare (say, once in 30 or 50 years) rainfall
total is likely to occur.

14.3 Climate and the Site

The climate of a site is the result of two factors,
the macroclimate and the microclimate. Macroclimatic information is available for nearly all sites
in which the architect is likely to be interested but
microclimatic information is sparse so that it is
necessary to apply the principles of micrometeorology in order to obtain some knowledge.

In the over-all climatic pattern, macroclimate is
dominant but microclimate can modify this appreciably. These major changes are brought about by
local relief, bodies of water, surface cover and adjacent constructions.

The local relief may alter the radiation load on a
building because of shading or reflection but its
main effect is likely to be on the air flow. Valleys

will tend to channel the wind into certain directions
as well as give rise to katabatic flow and frost hollows
(3.7.2). Exposed slopes or ridges will increase the
structural strain as well as the likelihood of dampness from increased winds. The air flow will naturally play a part in modifying the temperature and
humidity patterns, but its effect may be a subtle
one, such as transporting pollutants from distant
areas.

Proximity of the site to large bodies of water
should be noted since local land and sea breezes
(3.7.2) can occur and whilst temperatures are reduced during the day higher night-time temperatures are experienced.[17] The lake may also give the
ideal conditions for insect populations, such as
mosquitoes.

Surface cover to the windward can produce a
profound effect upon the climate; for example,
large stretches of tarmac, stone or concrete, such
as parking spaces or runways, give rise to high afternoon air temperatures which may influence the
adjacent site. Woods, grasslands, etc., will tend to
moderate the higher temperatures but may lead to
colder night minima and increased humidity.

Adjacent constructions may influence the site
climate in many ways. These can range from an
obvious shading of radiation to the numerous effects
of a city. The main differences between a city and
the local rural environments are given in Chapter
17.

14.4 Conditioning by Climate and Design

It has been described how the site of a building
may have its own climate, a climate that may be
rather undesirable. However, a logical approach
with respect to materials, design and positioning
may create a more desirable microclimate. Although
we are considering only individual buildings it is
possible to extend this idea of climatic conditioning
to include town planning aspects, such as parks,
street widths and orientations, etc.

In Section 14.7 the more general approaches to
climatic and design conditioning on the regional
scale are detailed. The question of house orientation on a particular site needs special consideration.
For instance, identical houses on opposite sides of
a street experience different house climates: one
may have a kitchen exposed to the midday sun
while the other receives the full impact of cold
north winds; one may have a breakfast room ideally

situated on the east whereas its mirror image across the street achieves its optimum temperature in the evening.

In the northern latitudes little radiation will fall on the northern walls while, in the temperate latitudes of the British Isles, the southern walls receive the most heat. In the equatorial regions where the sun moves through both the north and south sky during the course of the year there is very little difference in the loads received by walls of any orientation[18] (11.2). The light from the northern sky, being quite uniform in intensity, is desirable in laboratories and workshops. A recent construction at Leicester University has used the novel approach of angled windows to catch the northern light since it was not possible to orientate the laboratories north-south. In the subtropical regions the morning rays of the sun can be most pleasant but the evening rays, coming when the air temperature is high, can cause overheating and discomfort.

Concerning the air flow, it may not be desirable to have a long side of a house exposed to high winds, except for cooling purposes. For comfort and to reduce heat losses the entrances should be protected whenever possible.

14.5 Indoor Climate

In previous sections only the external climate of a building has been considered but to the occupants it is the indoor climate that is of the greater interest. It may be most unpleasant outside but if the intramural conditions are comfortable the architect has fulfilled at least part of his contract. The indoor climate depends upon many factors including the external climate, the materials of construction, the orientation, window size and aspect, ventilation and such artificial additions as heating and/or cooling.

Landsberg[19] quoted values of temperatures measured at noon on a clear day on both the outside and inside of walls. This experiment showed a 22 °C difference between the outside temperatures or radiated and non-radiated walls, a value reduced to 3 °C on the inside of the same walls. Schmidt[20] has reported that a brick wall of 10 cm thickness reduces the indoor diurnal range to about one-sixth of that on the external stone surface.

Related to this aspect is the delay in the times of extreme temperatures within the building compared to those outside. The lag is generally 2—4 hours[21] and results in the hottest period indoors

coinciding with the time when families are likely to be inside and cooking a meal, thereby raising the heat load.

If the occupant of a room is to be comfortable both the floor and the walls must be warm since radiative and conduction losses will otherwise be appreciable. Walls of high heat capacity are preferable as these retain a higher temperature during the night-time hours, a fact that is especially useful, economically, during long cold spells. Asphalt floors lose much more heat than concrete ones while a hardwood floor is a very good heat retainer.

Adequate ventilation is essential in all rooms otherwise the air will take up the stable thermal stratification with cold air at floor level and warmer air close to the ceiling.

When sleeping the human needs to be well protected from rapidly changing conditions of temperature and humidity for at this time metabolism is at a uniform low rate. It would appear from the few observations made that unless the temperature within the bed is about 30 °C untroubled, long sleep cannot be achieved. If the person enters a cold bed the equilibrium temperature is at the lower end of the range, the body normally raising the bed temperature by about 10 °C if the bed starts cold. The drop in bed temperature during the night is about 3 °C.

14.6 Heating and Cooling

It is when we begin to study these aspects that we realize it is quite impossible to consider all buildings together under one heading. We must subdivide into, at least:

(a) offices and buildings in which large numbers of people may be congregated;
(b) houses.

Until the recent energy crisis the tendency was to solve the climatic problems by building a large box, with or without windows, using artificial illumination and air conditioning to obtain the optimum conditions for the people within.

In large cities and for buildings in category (a) this method has an added advantage, for the lack of open windows reduces the pollution within the building, the people breathing only cleaned and purified air. Owing to the fact that by this means the workers can be given the optimum conditions for their labours, output is increased and the high cost of air conditioning can be more than offset. Even in the British Isles such methods are now being used. It is debatable whether or not there is a psychological disadvantage in working in a volume completely divorced from the external environment. In schools, for instance, distraction is reduced and as no more than 35 hours a week are spent within the building, the enclosed 'working ant' can at least be aware that an external world does still possess the diurnal and natural rhythms.

Although it may be agreed that, for a certain building at a certain site, heating and/or cooling plants are necessary, both the inital and the running costs can often be appreciably reduced by an application of micrometeorological principles. For example, the use of wind- or shelter-breaks can protect a house from the cooling effects of winter winds while trees can shade large areas of exposed wall and roof surface. In southern England very cold conditions are often associated with a predominantly easterly air flow so that a shelter belt of trees or fencing to the east of the house could prove most beneficial. In Texas, however, cold winds flow from the north and the belt would have to be of a different orientation.

The thresholds for working conditions vary with the labour and the people, different races being able to tolerate slightly different conditions, but it does appear that for the normally clothed human being the range between 18 °C and 24 °C is optimal, the best mental activity being achieved at the lower temperature, or even appreciably lower. At the low temperatures, physical discomfort can sometimes take precedence and cause lack of concentration in the mind.

In most areas of the world where scientific studies of heating and cooling are sufficiently advanced, it is generally considered that for temperatures below 18 °C some form of heating is necessary or desirable. It must be remembered that we are dealing here with the temperatures within a building, not the external shade temperature and there is little known correlation between the two, unless details of the construction are available. Thus, there is no general statement that can be made concerning the range of external temperatures that are optimum for the production of the optimum internal conditions.

For the category (b) it is now time to return to the concept of designing with the climate, a technique generally used until recent years. Sensible use of materials, ventilation, shading devices, and landscaping will help conserve energy. For example, windows set back or with overhangs cut out the high altitude sun but can admit the rays of the low winter, morning or evening sun. If this is so then it is possible to consider the range of external air temperature of about 13–28 °C as giving pleasant indoor conditions. The difficulty arises when dealing with housing in the densely populated urban districts since natural ventilation patterns are then vastly altered and there can be problems arising out of local laws concerning light admittance and house spacing. For houses the protections due to vegetation, such as shade trees and grass, reducing reflected radiation, should be utilized.

In order to estimate the fuel consumption of houses the concept of heating degree-days has been developed. This simple method uses the difference between the mean daily temperature, $\bar{T}_i$ (°C) and 18 °C, the daily value of the heating degree day being zero when $\bar{T}_i$ is greater than 18 °C – for no heating is then required. There is a very close correlation between fuel consumption and the heating degree-day value. It does tend to underestimate on very windy days, when the convection of heat is away from buildings and draughts are present. On very bright sunny days the method overestimates the amount of fuel consumption owing to the fact that the incident radiation is raising the temperatures of exposed buildings. A similar idea, named cooling degree-days, relates $(\bar{T}_i-18)$ °C to the energy needed to cool buildings to a comfort level. This expression is zero when $\bar{T}_i$ is below 18 °C, for then no cooling is needed.

In some parts of the world experimental houses have been constructed using solar energy to heat and/or cool the interior. Little information is available at present concerning the success and economy of these but the interested reader should contact the University of Arizona where research on this problem is being conducted. Another experimental plan is to use the soil both as a sink for heat dispersal and as a source for heat extraction.

Thom[23] has derived a rational method of determining summer weather design data that uses the distribution of dry and wet bulb temperatures. The techniques have been shown to give information of use to air-conditioning engineers, being an improvement over the use of dry bulb temperature alone. He has also derived expressions for use in calculating the heat system load from an extreme value analysis of dry bulb temperatures.[24]

Buettner[25] has suggested that in order to obtain an idea of the 'comfort cost' one should use the formula:

$$\text{cost} = \text{heating degree-days} \\ + 2 \text{ cooling degree-days}$$

and has constructed a map of the USA showing how this applies.

14.7 An Elementary Climatic Classification for Housing

14.7.1 *Tropics*

In this zone five main sections can be distinguished:

(1) the hot, humid   (4) the savanna
(2) tropical islands   (5) the uplands.
(3) the hot, dry

(1) In the hot, humid regions, the regions of minimum clothing (12.6), it is desirable to make the best use of whatever winds may be available and also to make certain that the building can radiate at night to the cool sky, while being sheltered at least partially from the day-time radiation.

In most of the hot, humid regions there is dense vegetation growth so that large clearings are made. The availability of wood allows the people to raise the houses for protection from wild animals, insect ravages and such like.

(2) The tropical islands follow much the same pattern as in (1) but here there are pronounced land and sea breezes so that the houses are designed for making full use of the cooling breezes. Often used are light bamboo or similar shutters that can be easily raised and then lowered during the times of intense rainfall. The buildings are usually just one room deep in order to take full advantage of the winds. Again the houses are raised for reasons given above, and also to obtain the benefit of the extra wind speed that occurs with increased height (Fig. 14.3).

(3) In the hot, dry regions, the semi-arid and desert areas, the radiation is so intense that protection has to be afforded from the sun's rays. Use is normally made of dried mud bricks as these give a good insulation. The houses are built of many storeys so as to catch the breezes and shade the

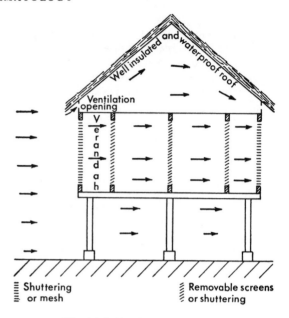

Shuttering or mesh   Removable screens or shuttering

Fig. 14.3 Hot humid zone house

lower levels. Generally the family adjourn to the flat rooftop at night to avail themselves of the radiative cooling and the cool winds. T. E. Lawrence (Lawrence of Arabia) wrote, concerning this aspect:

'Do not fall into the Khartoum fault of wide streets; in the Tropics air is an enemy, also sunlight.'

Design must be such as to keep out the heat of the day so that shades, small windows, white roofs and walls are an advantage. If possible it is desirable to screen the roof during the day but the balustrade around it must be fairly open so as to utilize the night breezes.

The small windows will act also as a protection when the sand and dust storms hit the area (Fig. 14.4).

(4) The savanna region combines the characteristics of (1) and (3), changing during the year. Here wood and brush are generally available but owing to the fact that large-sized logs, and other materials with large moisture content, split under the effect of the seasonal changes, dwellings are normally of mud and grass, erected under the protection of a shade tree and often surrounded by a thorn fence to protect them from the wild animals. In many places skins are also used in construction (Fig. 14.5).

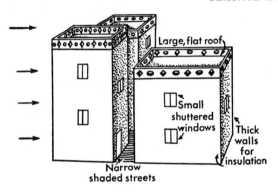

Fig. 14.4 Hot dry zone house

(5) The uplands are the region where, owing to the increased elevation, the temperature is reduced sufficiently so that there is very little discomfort felt as a result of high afternoon maxima. The problem here is to shut out the cold night air and buildings are therefore built with better insulation and protection from prevailing winds. Radiation to the to the night sky should be reduced as much as possible.

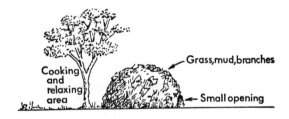

Fig. 14.5 Hot savanna zone house

14.7.2 *The warm or subtropical regions*

Here we may distinguish three main types:

(1) the Mediterranean climates
(2) the east coast regions
(3) the deserts.

In all these regions the solar heat is often very great with a high altitude sun and use should be made of shades, small penetration and white surfaces.

(1) In this region, supposedly one of the best areas for mankind, there is often much wood and workable stone. The region normally has hot summers and the stonework is used as good insulating material from the hot rays. There is much similarity

here to the hot, dry region of the tropics for the summer period is hot and dry. However, there is need to have more substantial protection from the cool winters and the rain that occurs then. Old dwellings in these regions often have open court-yards because the sun, which is never in the zenith, does not send any direct rays into these, yet they can radiate to the sky at night. This part of the house preserves a most equable climate and the coolness is often accentuated by the use of fountains or running water in the better appointed houses. It is interesting to note that in the northern Mediter-ranean area the roof as a gathering spot at night is supplanted by the less efficient balcony or pave-ment (Fig. 14.6).

Fig. 14.6 Mediterranean zone courtyard house (cross-section)

(2) The eastern coasts have a much larger rain-fall and the buildings have to be more substantial to give protection from the downpours. The intense radiation of the Mediterranean region summer is re-duced because of the summer precipitation but the extreme heat makes air conditioning desirable.

(3) These deserts are the regions of highest tem-perature on the earth's surface and are very sparsely covered with vegetation so that the nomad has to use animal skins to give him protection. Tents can be lifted at the side so as to gain the benefit of any breezes (Fig. 14.7).

14.7.3 *Cool regions*

Here the houses are normally small to conserve heat, there is plenty of wood and stone and in for-mer days use was made of thatch as a roof because of its good insulating properties. In this region, and the next, double-glazed windows and double doors will reduce conductive and draught loss. Excessive

Fig. 14.7 Hot desert zone tent

solar radiation is not a problem and houses should
be built to take advantage of available sunshine. If
bright sunlight is not required at certain times, cur-
taining can be used.

14.7.4 *Cold regions*

In the forested regions wooden homes are general, the roofs being steep so as to prevent a build-
up of quantities of snow (Fig. 14.8). Snow build-
up can present a real problem in cold regions for
a 10 cm depth of snow over a 40 m² roof creates an
extra 300 kg weight on the rafters. A paper by
Boyd[26] shows how climatological knowledge can
be used to compute maximum snow loads on roofs
in Canada.

The Eskimos have made the most sensible type
of house for the very cold areas. It is a circular house,
about 3 m in diameter, built of snow blocks or slabs
of stone, timber or whale ribs covered with walrus
hide. In the latter cases it is banked with snow or
earth for insulation. There is a slightly raised sleeping
platform covered with animal skins, the lowest skin
being set with the hair down to prevent the ice platform from melting. The entrance is tunnel-shaped,
below the sleeping platform level, and at either side,
within the house, are platforms for lamps. The heat
from these rises and adds to the body warmth of
the interior (Fig. 14.9).

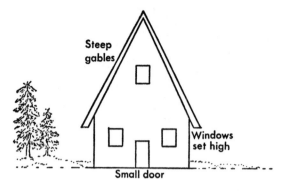

Fig. 14.8 Cold zone house

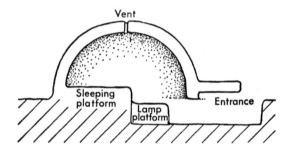

Fig. 14.9 Eskimo igloo

Some American Indians live in poor tepees
which are not suited to this type of climate. In
Siberia the natives occasionally use a round skin
tent with a smaller rectangular box of skins within
for sleeping quarters. The heating of the outer area
causes mildewing and the whole region begins to
smell, so that the skins have often to be replaced.

Chapter 15

CLIMATE AND HYDROLOGY

15.1 Introduction

This chapter deals with some of the problems arising from the role of meteorological phenomena in hydrology, the science concerned with the properties and other aspects of water. For the section with which we are concerned there has been developed the word hydrometeorology. The applied aspect of hydrology deals with such problems as flood control, irrigation and hydro-electric power.

The fundamental concept of hydrometeorology is the hydrologic cycle, the composite picture of the interchange of water and water vapor between the earth, the atmosphere and the seas.

The hydrologic cycle, see Fig. 15.1, has been examined in the USA over a long period and the findings of the Weather Bureau[1] are that, on average, over the whole of continental USA, some 65 cm of rain and 10 cm of snow are received per year. Of this amount 54 cm are lost through evaporation and transpiration, while 21 cm are lost by runoff and underground flow. At any one time the atmosphere is holding only about 2·5 cm of precipitable water so that the cycle must take place about once in 12 days. These figures are intended only to give an order of magnitude to the numbers involved.

Many of the factors included in the hydrologic cycle are, at present, difficult or impossible

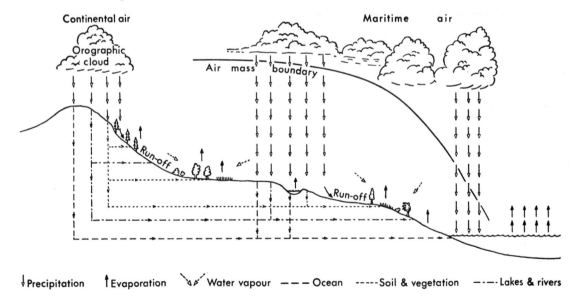

Fig. 15.1 The hydrologic cycle

to measure, such as ground water runoff, estimated through hydrograph charts, and evapotranspiration, estimated from equations or sparse measurements. However, if better use is to be made of water in the areas of the globe where it is at a premium for some time of the year, such areas being in the great majority, then a better understanding of the precipitation patterns must be acquired.

Hydrology has always been realized to be of great importance. Even the early civilizations appreciated their dependence upon adequate water supplies and records still exist of observations made concerning Nile floods as long ago as 3 000 B.C. The Pharaohs adjusted the taxes according to estimates of the height of the river. The Yellow river in China has also exerted a great influence on its people for thousands of years. However, knowledge concerning the pattern of the hydrologic cycle came much later, in about the fifteenth century.

15.2 Water Gain

Water gain to the earth's surface is received in the form of hydrometeors, such as rain, snow, hail, sleet, fog, dew and occult precipitation (see definition below). The measurement of some of these parameters has been discussed in Chapter 2 but it has been noted that the network of stations is not very dense. This lack of information means that we do not know with much exactitude the amount of water received at the surface, except as estimates over large areas and a long period of time. Our knowledge of the amount received in a small area in one storm, or during one day, is not very accurate. Lately attempts have been made using radar techniques to arrive at some estimate of the amount of rainfall received over a medium-sized area, usually about 150 km radius, during a period of time of the order of 1 hour. The techniques employed so far do not permit of accurate quantitative readings, although improvements are being made, but empirical techniques are allowing some estimates to be obtained and the subject is likely to show great improvements during the next few years.

Some of the terms used in describing precipitation are used very loosely. The meteorological definitions are given below:

fog — a cloud, with its base at the surface, con-

sisting of a visible aggregate of minute water droplets suspended in the atmosphere;

smog — a natural fog contaminated by industrial pollutants;

drizzle — very small, numerous and uniformly dispersed water drops, of a diameter of less than 0·5 mm; the droplets do fall towards the ground;

occult or horizontal precipitation — strictly a fog deposit on non-horizontal surfaces, especially vegetation;

rain — liquid water drops, diameter above 0·5 mm, that have a definite downward motion and reach the surface;

sleet — a mixture of rain and snow, also ice pellets;

ice pellets — transparent or translucent lumps of ice of diameter less than snow;

hail — balls or irregular lumps of ice of diameter 5 mm or more, produced by convective clouds;

snow — white or translucent ice crystals, usually of hexagonal form;

virga — water or ice particles falling from a cloud but not reaching the ground.

15.3 Catchments and Depth—Area—Duration Relationships

Precipitation is obviously the cause of all river flow so that little hydrological work concerning watershed catches or stream flow can proceed without some precipitation knowledge.

When the studies are concerned with long-term averages over large areas of homogeneous rainfall a thin network of stations will suffice to give a good correlation of river flow with weighted rainfall values. However, for most investigations dealing with short-periods, small areas and non-homogeneous rainfall patterns, a dense network of stations is necessary. The term homogeneous refers here to an area in which, over the time interval considered, most stations show similar patterns of rainfall distribution and variation.

Unfortunately, dense networks of stations are not common and the hydrologist has to make estimates of the water volume available during a period from a point sample. The need to introduce a volume/time factor has given the so-called depth — area—duration analysis much publicity.

Depth and area factors are not so easy to estimate as might at first be thought. First the area under study must be delimited and then, from observations, the amounts falling at rain-gauges within the area are plotted. Using the Thiessen network or the isohyetal method the average rainfall-depth over the area is then calculated. The two methods are shown by examples in Fig. 15.2.

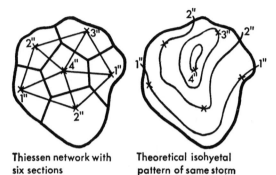

Thiessen network with six sections

Theoretical isohyetal pattern of same storm

Fig. 15.2 Thiessen and isohyetal methods

The Thiessen network, using perpendicular bisectors to lines connecting stations, delimits sections each of which is assumed to have a rainfall equivalent to that of the rain-gauge within the section. The isohyetal method can allow for orographic influences and other non-numerical information to be included.

A mass curve, accumulated rainfall against time, is used to determine the periods and amounts during the most intense part of a storm, the absolute maximum precipitation value. This value is considered to apply to a 40 km² area around the station. The technique enables depth–area–duration curves to be calculated and Fig. 15.3 gives

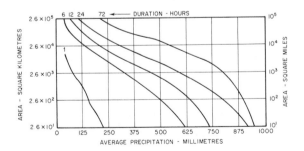

Fig. 15.3 Maximum depth–area–duration for major storms in the USA

some values found by the analysis of over 400 major stations in the USA.[2]

Interesting work in East Africa has shown that the stormflow prediction is best approached by use of a mean intensity weighted by the proportion of the total rain falling in each of several intensity classes. Runoff expressed as a linear relationship with rainfall gave a highly significant but not precise result.

Some idea of the intensity of rainfall over periods of less than 2 hours can be obtained by use of the equation:

$$i = a/(t + b),$$

where i is the average intensity over time t, and a, b are constants that vary with the region concerned. For periods of over 2 hours,

$$i = c/t^n$$

where c, n are constants that vary with the locality.

Values of a, b, c, n can often be mapped for a region so as to interpolate their values for localities having fewer data.

These equations can then be combined with Horton's depth–area relationship.

$$\bar{P}/P_0 = \exp(-kA^m),$$

where $\bar{P}$ is the average depth of rainfall for a given duration over area A, P_0 is the highest amount at the storm centre, k, m are constants for a given storm. The two equations will then yield a depth–area–duration expression.

15.4 Surface Runoff and Underground Water

There are three main paths that may be followed by the precipitation reaching the earth's surface before it arrives at a stream channel. These are surface runoff, subsurface flow and groundwater flow.

Surface runoff. Surface flow, sometimes called overland flow, is the water that travels overland to reach the channel. A channel here is meant to be any depression that may carry water during or after rain. The distance travelled in overland flow is generally small, with a few hundred feet as maximum, so the water soon reaches a channel and eventually a river. Hence, overland flow will soon give a peak flow to the river. The amount of surface flow naturally is reduced over permeable soil and, in general, surface runoff, except in urban areas, is only an important factor in streamflow due to heavy or high intensity rains.

Subsurface flow. This is also called interflow and is the water which infiltrates the soil surface, moving laterally until entering a stream channel. Because of the slower rate of movement it reaches the stream later than the surface runoff. In a uniformly[17] permeable soil the percolation tends to give little subsurface flow but where there is a thin soil over an impermeable surface there is a good interflow.

Groundwater flow. Also called base flow or dry weather flow. This does not fluctuate rapidly because it derives from the rainfall that percolates to the water table and it takes devious routes at low speeds. In some regions more than 2 years is required for the accretion of groundwater to be discharged into streams.

15.5 Engineering Hydrology

It is stated by Linsley[2] that 'the main purpose of engineering hydrology is to derive the factual information on quantities and rates of flow on which the design of engineering projects can be based'. Such a statement immediately emphasizes the role of meteorology in this field.

For the purposes of discussion we will consider the different problems under five main headings — storage reservoirs, flood control reservoirs, river forecasting, open channel hydraulics and river floods.

15.5.1 *Storage reservoirs*

Problems here depend upon the size of the reservoir. If the reservoir is small the knowledge of fluctuations in the daily rainfall and water demand may be necessary for full analysis. For a reservoir with very large capacity, of the order of several times the annual inflow, the drought period that must be analysed is of the order of years. In these cases annual totals of rainfall over the area may prove sufficient,[3] but monthly totals are needed for a comprehensive investigation. If the reservoir is of earth, high intensity rainfall may play a role in the weathering and wearing processes.

Of course, for all reservoirs, the evaporation problem is important and must be taken into consideration (see 3.8). It is an interesting thought that the two large man-made lakes, Nasser (in Egypt) and Mead (Nevada, USA), are in regions where evaporation is extremely high, resulting in the loss of tens of millions of litres of water per square kilometre each year.

15.5.2 *Flood control reservoirs*

This type of reservoir is designed to reduce the flood flow downstream to proportions manageable by the lower reservoirs. In general, for design purposes, this is often taken to be equal to the worst flood known to have occurred in the region of the reservoir. It must be borne in mind that, as mentioned by Linsley 'economic and social factors control the final design'.[2] Of course, allowance must be made for the advent of a second flood while the reservoir is still filled from the first flood, and this probability can be calculated if knowledge of the time distribution of rainfalls of flood intensity is available — assuming that it is known what classifies a rainfall of flood intensity.

15.5.3 *Forecasting river conditions*

The forecasting of river conditions is essential to all of the previous problems and this requires, among other factors, reliable information concerning present and future weather. For adequate use to be made of such information as rainfall, snow depth and density, temperatures, etc., good communications are essential. It is necessary for the hydrologist to be able to obtain the required data quickly from a wide area and thus a clearing centre for meteorological information can save him a great deal of time. Often the information of rainfall in daily amounts is insufficient for the hydrologist to produce a good forecast of river conditions and, ideally, rainfall recorders should be used.

15.5.4 *Open-channel hydraulics*

So far we have mentioned only those aspects of hydrology in which the static head of water plays the most important role. However, the hydrodynamic aspects of many problems are the most important, for instance the pressure of flowing water on structures. These extreme pressures occur when flood waves or bores are formed since these bring not only the increase in water pressure but an extra kinetic pressure due to the movement of large debris downstream. For a detailed account of open-channel hydraulics the reader is referred to Chow.[4] One aspect must be discussed in more detail: this is the problem of flood forecasting.

15.5.5 *River floods*

Floods occur when there is such a great discharge of water over part of the river course that certain aspects of the river vicinity are endangered; for instance, the banks break or overflow, dams are washed away or subjected to critical pressures and edifices on the river are destroyed. In general, the major meteorological factor will be the rainfall over a certain area during a certain time interval but floods can be caused by temperature-patterns. Such a relationship has been pointed out by Henry[5] to be the cause of floods on the Yukon river where the ice melts at the headwaters while downstream is still frozen. This then tends to become an ice jam that can dam the river completely.

Floods generally have three main causes: intense rains, the melting of accumulated snow or both factors together.

Some equations supposedly give the peak flow of a river as a function of the size of the drainage area, provided there is a hydrologically homogeneous region. The constants in the equations vary with the type of terrain, but the equations give no idea of frequency nor is a rainfall parameter contained in them.

Most equations do, however, use rainfall as one variable, and so it is necessary to know how often a rainfall of a certain amount is likely to occur within an area. For the USA maps are available,[6] based on calculations, to show the likely amounts to be received for a certain time interval (30 minutes, 1, 2, 3, 6, 12 and 24 hours) and a certain frequency period, such as every 1, 2, 5, 10, 25, 50 or 100 years. Analyses of this type are normally based on Gumbel's extreme-value techniques,[7,8] together with certain empirical relationships derived for the area under study.

Armed with these estimates it is then possible to use an equation, such as the 'rational formula'

$$q = ciA,$$

to calculate the flood flow, q. In this equation c is a runoff coefficient, i is the rainfall intensity over a duration equal to the concentration time of the basin and A is the catchment area. The coefficient c can be expressed as a formula relating it to frequency occurrence. There are many variations of these types of equation.

Chapter 16

THE URBAN ENVIRONMENT

16.1 Introduction

One of mankind's major problems is, of course, the rapid population increase. This in its turn leads to an even greater rise in urbanization, for people leave the land and the rural environment to seek work and, hopefully, better conditions within the industrial, city areas. It is estimated that in 1800 only some 2 per cent of the world's one thousand million inhabitants lived in cities of more than 100 000 persons. Today that figure is approaching 25 per cent and in some countries it is very much greater, for example, England and Wales (80 per cent) and the United States (70 per cent).

Detwyler and Marcus have expressed the role of the city extremely well:[1]

> Cities are the nodes of man's greatest impact on nature, the places where he has most altered the essential reserved of land, air, organisms, and water. The city is the quintessence of man's capacity to inaugurate and control changes in his habitat. Through urbanization man has created new ecosystems within which the interactions of man, his works, and nature are examples.

The impact of the 'new ecosystems' is certainly magnified in the cities, and a recent bibliography on 'Urban Modification of the Atmospheric and Hydrologic Environment'[2] cites hundreds of instances from around the world where the 'modifications' have been studied.

16.2 Physical Characteristics of the Urban Area

In this section the theme is the comparison between the urban and rural areas, in so far as the physical properties are concerned. It is these that then bring about the modification of the environment and the resultant 'city habitat'. In 61 A.D. the Roman philosopher Seneca noted 'As soon as I had gotten out of the heavy air of Rome and from the stink of the smoky chimneys thereof, which, being stirred forth whatever pestilential vapours and soot they had enclosed in them, I felt an alteration of my disposition.' In the mediaeval times when a large 'urban' area would have a population of only a few tens of thousands (London had about 50 000 population in the fourteenth century and was at least four times larger than any other English 'city'), the problems of pollution were appreciated. Bryant[3] writes 'The stink (of a town), too, was overwhelming, especially in the streets occupied by tanners and butchers. . . . Leading out of them were narrow alleys giving on to stables and lay-stalls and the foetid, tumbledown hovels of the poorer artisans and labourers.'

Anyone interested in the numerous atmospheric problems in urban areas should consult the World Meteorological Organization's, *Urban Climates*,[4] and the three volumes of preprints of the *Teaching the Teachers Colloquium* on 'Building Climatology'.[5]

The fundamental physical changes induced by a large city can be classified under three headings — hydrologic, thermal, and aerodynamic.

16.2.1 *Hydrologic changes*

In the countryside the rain falls, directly or indirectly, on soil of a permeable nature so that the amount of surface run-off is limited. In cities man generally succeeds in 'waterproofing', or making impervious, about 50 per cent of the surface. Streets, sidewalks, roofs, patios, and parking areas all increase substantially the run-off from even a slight rain. For example, from a 30 m x 30 m parking area a 1 cm rain will yield 9 m³ of water. This has to be transported away efficiently and rapidly otherwise serious flooding results. Even guttering on roofs[5] has to cope with large volumes of water to be fed either into the sewers or concentrated in some small area in the garden. An actual example of this 'waterproofing' occurred in the city of Kitimat in British Columbia. Observations made on the site led to the prediction that felling trees and paving the surfaces would lead to valley floods and, as a result, the valley bottom was not developed. This forecast was substantiated when, after a few years, floods of previously unsurpassed extent occurred.

Another alteration occurs in the out-flow of moisture via evaporation, for paved surfaces are dry for much longer periods than the neighbouring soils and vegetation so that less moisture is fed into the air. This has repercussions in three ways — less heat is used for evaporative processes and therefore remains to heat the surrounds, absolute humidity (the amount of moisture in the air) is low, and dust particles, being dry, are easily transported into the air by traffic and pedestrians (see 17.4).

16.2.2 *Thermal changes*

The city surfaces absorb appreciably more solar radiation than the corresponding rural surfaces, for a greater percentage of the reflected radiation is retained by the high walls and dark-coloured roofs. The concreted city surfaces have great thermal conductivity and capacity so that heat is stored during the day and released at night. The rural area, with grass or other vegetation acting as an insulating blanket, experiences a lower temperature during day and night that is enhanced by the evaporation and evapotranspiration taking place.

In addition a large amount of artificial heat is produced in the city. During winter in the mid and high latitudes the major source of heat in some cities is that produced within the city itself, and not the heat from the sun. A study of Manhattan suggests that the heat produced by combustion alone is $2\frac{1}{2}$ times greater than the solar heating. All heat produced in a house eventually

diffuses to the outdoors, automobiles add large quantities of heat and even the metabolic heat of persons is not inconsiderable.

In sub-tropical cities during the winter, heat production by man and his activities can be about 10 per cent of the solar radiation load.[6] In summer the percentage generally will be less but air conditioning can release an appreciable amount of heat at this period.

16.2.3 *Aerodynamic changes*

As far as the air flow near the surface is concerned, a city presents a very different profile from the country. In other words, the roughness of the surface has been altered. With increasing roughness the height of the wind layer affected also increases. One example showed that the wind speed reached 95 per cent of the free air speed at 200 m over level country, at 300 m over the suburbs and at 400 m over the city. Investigations suggest that the roughness increases with the width of the building and the square of its height, but is inversely proportional to the size of the lot occupied by the building.

This aspect of a reduced wind speed tends to cause pollution concentrations and increase stagnation periods. However, within the confines of the major downtown area a pronounced channelling of air flow and an increase in the turbulence (eddying) occurs so that at certain locations air speed may actually be greater than that in the open countryside.

Studies of air flow in cities are few and non-standardized, but it is suggested that the air speed on the leeward sidewalk is often only about half that on the windward side. Trees along the roadsides also play an important role in reducing wind-speeds. A study in Frankfurt found that wind speeds in the streets were related linearly to those measured at roof-top level and that wind speeds on the roofs of the windward side of the street were some 30 per cent greater than those over roofs on the lee side.

16.3 The Climate of the City

A collection of the average changes in the climatic elements caused by the city is given in Table 16.1, based on Landsberg.[4] The major reasons for the changes have been given in 16.2

TABLE 16.1

Element	Parameter	Urban compared with rural (−, less; or +, more)
Radiation	On horizontal surface	−15 per cent
	Ultraviolet	−30 per cent (winter); −5 per cent (summer)
Temperature	Annual mean	+0·7 °C
	Winter maximum	+1·5 °C
	Length of freeze-free season	+2 to 3 weeks (possible)
Wind speed	Annual mean	−20 to −30 per cent
	Extreme gusts	−10 to −20 per cent
	Frequency of calms	+5−20 per cent
Relative humidity	Annual mean	−6 per cent
	Seasonal mean	−2 per cent (winter); −8 per cent (summer)
Cloudiness	Cloud frequency and amount	+5−10 per cent
	Fogs	+100 per cent (winter); +30 per cent (summer)
Precipitation	Amounts	+5−10 per cent
	Days (with less than 5 mm)	+10 per cent
	Snow days	−14 per cent

but will be discussed in more detail in this section. Two special terms have developed related to city climates — the urban heat island and the dust zone.

16.3.1 *The urban heat island*

It was explained in 16.2.2 that, due to the physical properties of the city, temperatures would tend to be greater there than in the surrounding countryside — a fact confirmed in Table 16.1. Recent studies in Columbia, Maryland, have shown that even one block of buildings start the heat island formation. Meso- and micro-climatic variations due to special conditions, such as frost hollows, can give rise to great differences — a change of 21 °C is cited for Madison, Wisconsin. Interesting increases of 1−2 °C have been noted at traffic lights due to cars idling there.

On clear days highways and asphalt surfaces can average 10−20 °C or more above those of forests or graasy areas. Even on overcast days buildings show a 3 °C increase above the surroundings.

The extra heat can reduce the heating degree days total by 5−15 per cent but can cause an increase in the number of cooling degree-days. The frost-free season is often appreciably greater within a city than its surroundings — at Munich, Germany, this difference is as much as 60 days,

but values of 1−3 weeks are more normal. This extra heat in the city gives rise to a low pressure area and an inflow of air from the surrounding countryside. It is suggested that the critical wind speed for the destruction of a heat island is related logarithmically to the population count of the city.[4]

16.3.2 *The dust dome*

As mentioned earlier, much dust is generated within a city and it can be seen as a 'dome' over most cities as one approaches by air. The dome becomes especially well defined on a relatively calm day with a natural circulation such as depicted in Fig. 16.1. On these days the recirculating system continues to pick up particles and acts towards concentrating particulates in the city. Large dust particles (diameter over 5 μm) fall out rapidly but others remain in suspension and can act in the condensation processes. Gaseous pollutants may follow a similar pattern but, unless trapped beneath an inversion, do not sink and recirculate.

The dust, smoke and other particulates combine to give the air a certain turbidity, but all air is turbid to some extent. This dust has a pronounced effect upon the shorter wavelengths and reduces both ultra-violet intensity and sunlight by substantial amounts (Table 16.1). The

Fig. 16.1 Schematic of the local air circulation, differential cloudiness, and shape of the 'haze hood' above an idealized city. After Lowry.[7]

decrease is greater in winter than in summer due to the increased path length of the solar beam through the more turbid atmosphere.

The above combination naturally will reduce visibility within the city and increase the frequence of fog. The fog incidence is magnified due to the condensation of water vapour on the many dust particles.

16.3.3 *Precipitation*

The previous sections have indicated that human activities produce a great number of particles, condensation nuclei, in and around cities. This fact very likely contributes greatly to the increased dampness and precipitation noted over cities. In many cases the prevailing wind causes the maximum rainfall area to be moved downwind of the city centre or industrial region. In some cities there is an appreciable increase in amount much in excess of the 5—10 per cent given in Table 16.1. Investigations suggest that the increase is greater in the winter and that both snow amounts and thunderstorm frequency are also increased.

A study of thunderstorms in the London area showed that they yielded 30 per cent more rain over the city than over the surrounding countryside, but a classical case gives a thunderstorm yielding 68 mm over the city and less than 3 mm elsewhere.[4] This could be due to a combination of the dust and particulates plus the increased convection due to the heat island.

16.3.4 *Small heat variations*

Within the city large climatic variations can exist within short distances. Landsberg[4] has given details of temperature variations in specific locations in Columbia, Maryland. During a summer afternoon temperatures over a lawn and a

courtyard were both found to be 87 °F (31 °C) compared with 111 °F (44 °C) over a parking lot. By two hours after sunset the courtyard surface was 3·6—9 °F (2—5 °C) above the air temperature, itself 3·6 °F (2 °C) above the grass surface temperature.

Even small parks can create a modification, as proved by investigations in a small park (0·84 acres, 0·34 hectares in area) in Cincinnati.[1] During the afternoon the park area was about 2·7 °F (1·6 °C) cooler with a significantly lower heat load in a ratio of about 3:7 compared to the region away from the park. Rivers and small lakes or ponds show lower temperatures on summer evenings than the neighbouring built-up areas. The new technique of infra-red thermometry is particularly useful in identifying relatively 'cool' and 'hot' regions.

16.4 Pollution in the City

Industrial areas are a feature of most cities, and their effluent, together with the concentration of automobiles, heating/cooling devices, and the city 'dust' mean that the urban region is a great source of pollutants. Previous sections have shown how the city tends to develop its own, often closed, circulation cell thus causing a concentration of pollution to occur. It is particularly important to note that the 'urban cell' extends beyond the city boundaries for the effluent from an industrial region outside the city may be drawn in to the 'cell'. Another fundamental aspect is that growth of the city and the corresponding growth of the 'cell' may then embrace an industrial region and increase city pollution. The desirability of a green belt is clear.

There is a large increase in the amount of dust in city air compared to country air (4—1 000 times as great). It has been found that an appreciable part of this dust is lead from the automobile exhausts. Dust, like the automobile gases of CO, NO_x, is normally generated over large regions and is referred to as an area source. The concentration, for the city as a whole, is often a function of the meteorological conditions. The atmospheric conditions can exaggerate this concentration under certain criteria, of which two characteristics are of the greatest importance — air flow and temperature stratification. If the wind speed is great the pollution will be mixed, or diluted, with a larger volume of air than under lower wind speeds.

If the thermal stratification is such as to prevent air from mixing vertically, such as when an inversion occurs, the potential volume of air available for dilution is further reduced. Therefore, if an area can have warning of when conditions of low wind speed and inversion formation will occur steps could be taken to reduce pollutant emissions and prevent or lessen the pollution episode. Forecasts of these 'stagnation' episodes can be issued by the Weather Service and are often associated with the occurrence of anticyclones, or high pressure cells, which often have characteristics of lighter wind speeds and clear, dry air, ideal for inversion formation. Cyclones, the faster moving low pressure cells, will tend to ventilate the areas. However, other conditions, such as stagnating fronts, can help to entrap and concentrate pollutants.

This pattern is very different from that given by point sources, such as industrial chimney stacks, for then the location of the source plays a dominant role and the concentration generally decreases rapidly, but not uniformly, with distance and direction of the source.

A particular pollutant problem that relates to atmospheric conditions concerns photochemical smog. When solar radiation is high the insolation reacts with certain hydrocarbons and oxides of nitrogen (mainly from car exhausts) to form photochemical smog, a phenomenon that irritates the eyes and mucous membranes, injures plants and reduces visibility.

16.5 Hydrology in the City

As with population pressure and pollution the urban environment tends to magnify the problems of water drainage. In general, urbanization both increases the magnitude of floods and their frequency. The root cause is the 'waterproofing' referred to in 16.2.1 in conjunction with man-made changes to the natural topography — such as filling in of channels. In addition urbanization alters sedimentation rates in rivers, water quality and water temperature.

It appears logical to mention the above in a book on climatology because the water problems are stemming indirectly from a climatic element, precipitation. Perhaps more directly, however, we should consider the 'natural atmospheric hazards' — and drought, hurricanes and winter storms are excellent examples.

Drought, or a period of abnormally low precipitation, can put a great stress upon a city's water supply, especially when citizens continue to try to keep lawns, shrubs, and gardens green. However, industry is often the greatest user of water and severe shortage can cause production cutbacks and lay-off of workers with resultant economic stresses.

Hurricanes, generally occurring at maximum intensity within about 100—200 km of the Gulf and North Atlantic coasts, bring significant rain, flood and wind hazards in the period of July to October. These tropically-spawned storms can be associated with heavy rainfall (500 mm in 1—3 days), adding extra problems to the wind and, sometimes, sea surge difficulties experienced in the city.

Winter storms cause a particular problem in cities for heavy snowfall or glaze ice can disrupt urban communication and transport in a very short time. Even areas that are prepared for heavy winter snows can be caught out if a storm occurs early, before snow ploughs and other equipment are ready. As with flood precautions too much attention is focused on 'average' conditions.

Chapter 17

THE CHANGING CLIMATE

17.1 Introduction

In many recent journals, magazines and newspapers we have been made aware of the subject of climatic change. This has become, like 'environment' and 'energy' a scientific subject of great public interest. The intention in this chapter is to study the facts of the case — to survey what we really know.

As defined earlier, climate is concerned with long-period manifestations of the weather or the composite weather conditions over periods of years. There is, unfortunately, an appreciable degree of confusion among both scientists and the public between the terms 'weather' and 'climate' simply because the definition, while concerned with time, does not — and cannot — give a temporal breakpoint at which weather becomes climate. For convenience a period of 30 years has been adopted as the time over which climatological normals are calculated — it could as well have been 50 years or 19 years (as in the calculation of mean sea levels). This concept of the 'climatological normal' has led many persons, including some scientists, subconsciously to view the climate as 'fixed', 'unchanging', except over geological time scales to which we shall refer later.

Because weather is the result of many interplays and interrelationships among numerous energy sources and sinks it is *not* a settled phenomenon. The atmosphere—water environment may strive for internal equilibrium but it is never attained. The atmosphere is changing on all time and volume scales — *weather is never static; climate is never static.* As can be seen from Fig. 17.1 there is variation of temperature on all time scales and even year-to-year changes in, for example, rainfall can be of large magnitude (Fig. 17.2).

By these definitions it is clearly illogical to speak of a 'climate change' over a period of a year or even a few years. It has been suggested that for periods of from 1 to 100 years we should speak of climatic fluctuations, reserving the word 'changes' for longer periods. The rationale for this could be that since, for extra-tropical stations, the mean annual temperature has a standard deviation (a measure of variation) of about $1°C$ a stability or possible error of $0.1°C$ in the climatic normal will require 10^2 or 100 years.

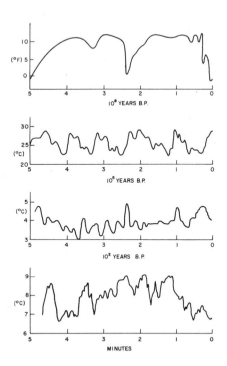

Fig. 17.1 Temperature variations on various time scales. B.P. = before present.

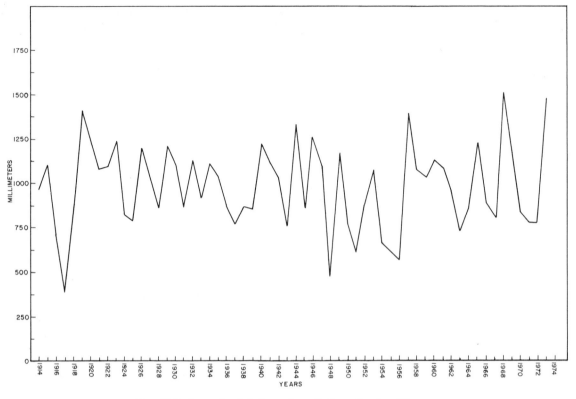

Fig. 17.2 Annual rainfall totals at College Station, Texas

17.2 Climatic Factors

If we are to consider climatic changes it is essential to understand what are the causal parameters (factors) of climate. In this manner we may concentrate on any possible changes in these factors while realizing that the climate of necessity then will alter.

There are only four basic factors of climate — solar radiation, the underlying (or radiation receiving) surface, the earth's rotation, topography and the land/sea configuration. These four act, via the secondary factors of pressure, air flow, ocean movements and air masses, to determine the climate.

However, before we can appreciate present climatic patterns and try to estimate future conditions it is necessary to gain some awareness of past climates. As Lord Byron remarked 'The best prophet of the future is the past.'[1]

17.3 Methods of Assessing Past Climates

There are three basic ways in which past climates have been assessed:

(1) using the uniformitarian principle that the response of an organism to a certain climate has been the same over all time,
(2) observations of geological forms or deposits,
(3) calculations using relatively simple approaches.

17.3.1 *The uniformitarian principle*

This method has been applied to fossil evidence of flora and fauna. With regard to the latter, mammals, being homeothermic, are less susceptible to thermal changes than the poikilothermic reptiles. In a related technique the ratio of the two oxygen isotopes (oxygen-18 and oxygen-16) present in the fossil is measured. This ratio has been shown to be related to the temperature of the water in which the invertebrate animal lived.[2] Another approach uses the Globigerina pachyderma,[3] which spirals to left or right, dependent upon water temperature.

17.3.2 *Sedimentary or geomorphological evidence*

The use of landforms, and especially glacial features, has often been involved in making climatic inferences of the past. Also, lake and sea level fluctuations have been interpreted climatically, and, in addition, weathering patterns have been used as evidence. Another example uses salt deposits, for these can be formed when, over long periods, evaporation exceeds precipitation; while aeolian sands can indicate ancient deserts. A particularly useful approach has been the study of varves,[3] the deposits on the floor of marginal glacial lakes. During a warm (summer) period a light coloured layer of silt settles from suspension when surplus water from ice and snow melt yields a thick layer of finer, often organic, material is laid down. The constitution and colour of these gives an indication of the temperature patterns, and the technique has been particularly useful in Scandinavia and Alaska.

17.3.3 *Calculations*

The most realistic method under this heading uses orbital fluctuation. The orbit of the earth around the sun is subject to certain variations. Some of the variations of importance in estimating the incoming solar radiation were studied in detail by Milankovitch. These were:

(1) the obliquity of the ecliptic (tilt of the earth's axis) that varies over 40 000 years from about $21.6°$ to $24.6°$ (presently $23.45°$), with the larger obliquity causing greater seasonal contrasts; and

(2) the precession of the equinoxes (changes in the time of closest approach of the earth to the sun, now on January 3rd) that has a 21 000 year cycle.

Using this information Milankovitch[3] calculated radiation inputs for various latitudes in the northern and southern hemispheres for the past 600 000 years. However, the relative weighting that he gave to the two items has been challenged and recently other patterns giving better correspondence between the curves and known Ice Ages have been devised. An interesting facet is that the patterns for the two hemispheres or comparative latitudes are not coincident. Many years ago, Brooks,[4] an English meteorologist, used other methods such as the elevation of the land, land/sea configuration, volcanic action and ocean currents to calculate the existing climates during certain epochs. From this technique the first temperature curve in Fig. 17.1 was deduced.

17.4 The Historical Past

In the historical past, from about 4 000 B.C. to, say, 1 750 A.D., various means have been used to assess climates. These methods have included

(a) records of floods and droughts – such as those kept in the Nile Valley;

(b) large-scale out-migration from areas – for example, of the Vikings or the Mongols;

(c) contemporary literature – chronicles, diaries, paintings;

(d) agricultural and constructional evidence – old irrigation and water storage systems, such as those in Aden that can hold millions of gallons, in an area that now receives less than 50 mm per year;

(e) palynology – the study of pollen grains or spores; a layered sequence of pollen in ancient lakes or peat bogs is studied by core sampling;

(f) dendrochronology – tree-ring investigations, uses the fact that the growth rings can record significant climatic events – unfortunately, factors other than climate can affect the thickness of the rings.

17.5 The Present

During the past two centuries instruments have developed so that regular, reliable readings can be obtained, although not for all areas and all times. In fact, for example, only a handful of climatic stations have records extending over 200 years that can be judged reliable, representative, and not subject to man-made change – such as the growth of a city. This period is referred to as the secular or instrumental period.

17.6 Deductions Concerning Past Climates

The impetus for assuming climates have not always been as they were occurred in 1821 in Switzerland when an engineer, Venitz, used geomorphological evidence to suggest that Alpine glaciers were remainders of very much larger fields. Since that time many studies have focused attention on certain geological periods and places. Other studies, based on Wegener's work using continental drift, derive patterns in various epochs.

In the Holocene (the past 10 000 years) the following 'spells' have been differentiated:[3]

4 000–3 000 B.C.	The climatic optimum — mean temperature of the middle latitudes 2.5 °C above present, heavy rainfall in Nile basin.
3 000–750 B.C.	The subboreal phase — Europe cooler, moister, less rain in the Nile area.
750 B.C.–800 A.D.	The 'classical' — milder, much rain over Mediterranean, very wet in Nile area and Yucatan.
800–1 200 A.D.	'Little climatic optimum' — warmer and drier in North-West Europe and North America.
1 200–1 500 A.D.	'Mediaeval climate' — cooler, stormier, winters extremely cold, dry spell in South-West America in thirteenth century.
1 550–1 800 A.D.	'Little Ice Age' — mean temperatures in northern hemisphere, 1 °C below present.

A more detailed table is given in *Physical Climatology*.[4]

17.7 Effect of Changes in Causal Parameters

The basic factors of climate have been shown to be solar radiation, underlying surface, earth's rotation, topography. Solar radiation is in turn affected by the earth–sun geometry and the atmospheric components while the underlying surface can be changed by nature or by man. The latter two factors are, at present, affected appreciably only by other than man-made forces and efforts and may be assumed constant.

17.7.1 *Solar radiation*

The amount of radiation received at the outer surface of the earth's atmosphere is related to the solar constant, a standard energy measurement used in meteorology. The absolute value of this 'constant' is not known and is said to vary irregularly by about ±1·5 per cent. It is interesting to note that all the climatic variations of the past could be explained by fluctuations of as little as ±8 per cent in the solar

constant. Budyko[5] has estimated that a 1 per cent change in radiation would lead to a 1·2–1·5 °C change in the mean annual temperature of the earth. A simple calculation shows that, if the thermally active layer of the sea is assumed to be 100 m deep, a storage of only 0·7 langleys/day† (a cloudless day in summer at mid-latitudes has about one thousand times this amount) would produce a temperature increase of 0·3 °C in ten years.[6] To monitor such a radiation change we should need to measure the solar constant with 0·1 per cent tolerance. Satellite infra-red scanners may be able to give us a clue as to the water temperature changes when the techniques are refined but currently we do not know the ocean temperatures over large areas with extreme accuracy.

While on the aspect of ocean temperatures we must mention the two important facts that (1) the water-atmosphere energy systems are interlinked and (2) the heat capacity of water is about 3 000 times that of air. In other words, the energy released from a metre of water cooling by 0·1 °C would heat 30 m of air by 10 °C. In addition there is about a 300:1 ratio between the mass of water to that of air. Thus, the air has a short 'memory' while the oceans have a long 'memory'. Perhaps some perturbation occurring in the abyssal depths 1 000 years ago could cause a gradual change in air temperatures now. On average the heat flow is about 1/3 000 of the solar radiation input but in some volcanic regions the heat flow is one thousand times greater, while for volcanoes in action (that may be below the oceans) it may be 10^5 or more times greater.[7]

17.7.2 *Atmospheric composition*

The major constituents that affect the radiation received at the earth's surface are dust, carbon dioxide and water vapour.

Dust — It has been suggested that an increase of two million tons in the atmospheric loading would be capable of reducing the world mean temperature by 0·4 °C.[8] Remembering this it is intriguing to note that:

(a) a big volcanic eruption, such as Tamboroa, Krakatoa or Agung, adds 100 million tons of dust to the atmosphere;

(b) man-generated particulate input is about 500

† 1 langley = 1 cal cm^{-2};

million tons per year, of which about 1 per cent, or 5 million tons, remain in the atmosphere for long periods.[6]

While on the subject of dust we should recollect that dust fall on glaciers and snow will lead to increased radiative absorption and faster melting while the particles themselves can act as rainfall nucleii and lead to increased precipitation. Mitchell has suggested that dust particles in the lower atmosphere tend to cause a local heating but in the upper atmosphere they can lead to an overall cooling effect. Particles of diameters greater than about 5 μm soon settle but smaller motes can remain in suspension, unless washed out by rain.

Carbon dioxide – Carbon dioxide, currently at a concentration of 320 parts per million, increases at about 1 p.p.m. annually, an increase that should result in a temperature increase of $0.01\ ^\circ$C each year.[9] Such an increase has not been obvious during the past 20 to 30 years. Another implication of the carbon dioxide increase is discussed later.

Water vapour – Changes in water vapour in the atmosphere, over long periods of time and large areas, are unlikely to be great enough to be important. However, water injected at high altitudes, for instance by jet aircraft, can cause cirrus formation (contrails) and thereby reduce radiation input.[10]

An overall aspect in which the atmosphere plays a role is in the reflectivity of the solar radiation by the earth-atmosphere system, estimated as 34 per cent. Recent calculations suggest that if this percentage changed by a single point the magnitude would be sufficient to account for the temperature drop believed to have occurred in the past quarter-century.[6] *If the temperature drops then ice and snow coverage increases,* the reflectivity increases and the temperature falls – a self-strengthening facet of the complex energy interplays – rather like the collection of snow around a stone rolling down a slope. It is a subject of discussion among climatologists as to whether the present climate is in a stable or unstable stage.

17.8 The Greenhouse Effect

The atmosphere absorbs selectively in such a way as to transmit most of the solar radiation but to absorb a large percentage of the terrestrial, or long-wave, radiation. This tends to conserve heat energy in the earth–atmosphere system and is referred to as the 'greenhouse effect', because of the similarity of the atmosphere to the glass, in so far as the electro-magnetic characteristics of transmission are concerned.

If, due to changes in the atmospheric transmissivity, the magnitude of the greenhouse effect is altered by 1 per cent the earth's surface temperature would change by about $0.3\ ^\circ$C and the average vertical lapse rate of temperature by some 0.8 per cent.[6]

17.9 The Vertical Lapse Rate of Temperature

Because of the three dimensional character of the air–water system one cannot focus on the surface characteristics alone. The vertical lapse rate of temperature has been shown by Smagorinsky[6] to be related to some basic flow patterns in the atmosphere, such as the position of the semi-permanent subtropical anticyclones.

A change of 1 per cent in this parameter can cause a location change of about 0.2° latitude in the position of these anticyclones. In addition, past records suggest that a change in the location of these anticyclones of x° latitude is related to a $3x^\circ$ latitude shift in the inter-tropical convergence zone (I.T.C.Z.) over West Africa. Here the I.T.C.Z. is the region between the dry, desert air and the moist, maritime air. Thus, a change of 0.2° in the anticyclones is associated with a 0.6° latitude relocation of the I.T.C.Z. over West Africa.

Because of the large N–S gradient of annual rainfall over West Africa a shift of 0.6° southward is equivalent to a decrease of about 100 mm of rainfall per year. Similar calculations, using the change in the lapse rate caused by the carbon dioxide increases estimate a decrease of 86 mm per year.[6]

17.10 Changes that may be occurring now

Evidence for significant changes has to be rather circumstantial by its nature. Clearly, a trend may cease and reverse itself for reasons that, at present, are imperfectly understood. Let us consider the types of simple change that may occur.

In this contrived graph (Fig. 17.3), which resembles many patterns of mean annual temperatures, one hundred years of record is composed of

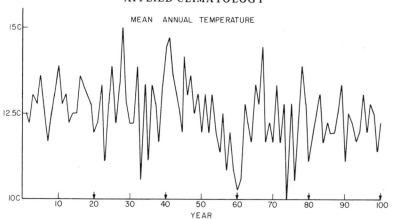

Fig. 17.3 A contrived graph of mean annual temperatures

five distinct twenty-year periods. In the first, the base condition, the mean and the standard deviation are 12·5 °C and unity, respectively. In the second twenty years the deviation has doubled, in the third period there is a downward trend so as to make the mean decrease to 12·25 °C. In the fourth spell the mean is now 12 °C and, finally, the mean remains at 12 °C while the deviation returns to unity. Such a pattern, which *may* occur, could not be identified statistically unless one knew in advance what to search for.

There are two basic approaches used to 'identify' changes, observational and theoretical. Let us consider these in turn.

17.10.1 *The observational approach*

Analyses of instrumental observations have been used by various researchers to show different things. Although some climatologists talk freely of 'the cooling since the 1940s', I can find no detailed, comprehensive world-wide study that categorically confirms this statement. These climatologists are generally citing the work of the two leading investigators, Mitchell and Budyko. Mitchell's classical paper[11] uses nearly 180 stations prior to 1940 and only 122 since, a sample change that, together with a removal of an unknown 'urban heating' trend, could induce an incorrect interpretation. As with Budyko's anomaly technique[5] (Fig. 17.4), using areas and not individual stations, there is insufficient data given in the paper to verify the analyses. Callendar[12] states that Japan in 1948–57 had the warmest decade since 1880, Europe had very warm

summers in 1957 to 1959 while the Canadian prairies and the adjoining plains of the USA had mean temperatures in 1952 to 1958 at least 0·6 °C above the 1901 to 1930 average. Dorf, an American paleobotanist, in 1960 stated '. . . we are apparently well along in a general world-wide warming trend' and also 'for the past one million years the earth has been subjected to an abnormally cold climate.'[1] It is a sobering experience to study ocean levels for these show an increase through the mid-1960s in latitudes up to about 60 °N and a slight decrease polewards.

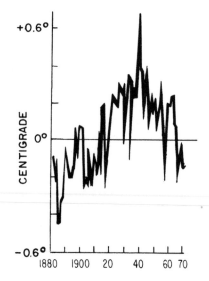

Fig. 17.4 Budyko's pattern of temperature changes in the northern hemisphere

This suggests melting of glaciers as a cause – hardly in accord with a cooling trend.

Bryson, the recent promoter of the 'cooling concept', uses data concerning ice conditions around Iceland to make his point. However, Lamb, an equally renowned British climatologist, suggests the decrease is not so drastic and cites values that are not in good agreement with Bryson's presentation.

Although temperature is the most popular element in climatic change studies, rainfall data have also been used. Kraus suggests that tropical rainfall decreased abruptly at the end of the nineteenth century,[14] a belief not held by many other researchers. It must be remembered that the amount of water an air mass can hold is dependent upon its temperature – an 11 °C rise will approximately double the saturation water vapour pressure. Thus, a cooling may well be associated with a drier period, but not inevitably so.

It is clear that observation cannot 'prove' that a change in any element, even if its significance is validated statistically, will continue even for one more year. The classical example of 'unsafe' extrapolation is shown in Fig. 17.5. From 1890 until about 1930 there was a high correlation between the level of Lake Victoria and sunspots.[15] From 1930 the correlation has been negligible and useless for forecasting purposes.

How complicated the picture really is has been stated by Budyko: 'Climatic changes during the period of instrumental observations have not been uniform for various parts of the earth and for different seasons of the year',[5] while Lamb comments 'there is no contemporaneity of climatic change.'[16]

17.10.2 *The theoretical approach*

In an earlier section it was mentioned how specific changes in the causal parameters, if occurring, would affect certain climatic patterns. However, the only two aspects that we can say, with some conviction, are really changing are the carbon dioxide and the dust content of the atmosphere. Perhaps these *are* sufficient to induce a climatic change. However, Dyer suggests that there has been no convincing evidence for a recent world-wide decrease in atmospheric transmission.

It has been a favourite pastime of many investigators to look for cycles of approximately eleven years in climatic phenomena so that it may be related to sun spots – as in the Lake Victoria example. It must be noted, however, that although over the past 250 years the sunspot periodicity has averaged

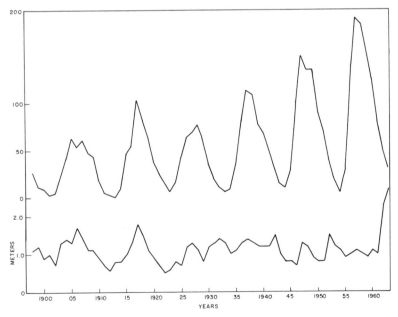

Fig. 17.5 The variations of sunspot number and the level of Lake Victoria

11·2 years, the cycle fluctuates between 7·5 and 16 years and the amplitude varies by ± 50 per cent. Lamb has suggested a cycle of something less than 200 years in European climate: could this be linked with the 179-year periodicity of the Grand Alignment with its apparent effect on sunspot intensity? A number of years ago Brooks wrote 'The literature concerning sunspots and their influence on weather is extensive, complex and generally vague.'[17] Such an observation is equally true today.

An especially important facet to watch is the behaviour of the circumpolar vortex, a flow of high altitude westerly winds related to the transition zone from tropical to polar air. There is good evidence that the lower edge of this vortex has remained further south during the northern summer in recent years and theory suggests this will have important repercussions. This may be related to changes in the surface temperature of large areas in the Pacific — but what causes these and are they forecastable? If these water temperature changes modified typhoons and hurricanes, the meridional transport of heat would be altered also.

17.11 How would Climatic Change affect us?

A question often posed is 'How large must a climatic change be to become important?'. This could be rephrased to ask 'How large must a climatic change be before it affects us?'. Basically, any climatic change or fluctuation is bad since life (people, animals, crops, vegetation, rivers) is adapted to certain sets of conditions.

A climatic change does not have to be large, in absolute value, to be important. For example, a decrease of 0·6 °C in the April to October temperatures would alter the growing season of N.W. Europe by about ten days and the growing degree days (related somewhat to plant growth) by 15 per cent. In addition such a decrease would lower the agriculturally productive area by about 100 m on hill slopes.

Considering a small change over a much smaller area than N.W. Europe, the next figure is rather thought-provoking (Fig. 17.6). This shows that, dependent upon the level of the critical temperature for the problem on hand, a change of even 0·1 °C could change a growing season by 6 to 7 days. A thermal change that would be hard to identify by observational methods.

Again, a decrease of 1 °C in winter temperatures would increase fuel consumption by some 10 per

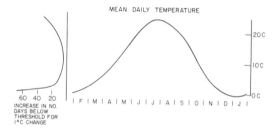

Fig 17.6 The change in growing season with temperate changes

cent in the Gulf coast area and by 3–4 per cent in the northern plains states. Of course, in some areas such as the tropical islands a decrease of 1 °C, year round, would be welcome and hardly noticeable. Thus, we see that climatic changes of equal magnitude may have different practical impacts in different areas.

17.12 The Future

It has been said that the human race is poised precariously on a thin climatic knife-edge where only 30 °C represents the temperature difference between the tropics and the poles. In addition the average global temperature is only 8 °C warmer now than during the last Ice Age, during which thick ice covered large portions of North America and Europe.

The importance of conserving our atmospheric environment is well illustrated by the following example. If the transmission coefficient of the atmosphere decreases for any reason by about 5 per cent the solar radiation received at the Equator would be reduced by 6 per cent, at 30° latitude by 8 per cent, and at 60° by 10 per cent. Furthermore, at 60° the summer reduction would be 9 per cent compared to a winter reduction of 16 per cent! This very simple calculation shows how atmospheric changes can affect some seasons more than others — proving Lamb's point of the non-contemporaneity of a standard climatic change.

Perhaps if the climate is balanced delicately, a small change in some facet may have far-reaching effects while an equal change in another parameter may simply induce perturbations that soon will be damped out. The meteorologist's primary problem at present is to analyse the climatic behaviour and strive, through physical considerations, accurate and wide-spread measurements, and modelling techniques, to extract the cause—effect relationships. Currently we cannot say that definite trends or long-period regular cycles exist but we are aware that variations are the 'norm'. The magnitude of these fluctuations is known only approximately but these should be considered in long-range planning, especially for agriculture. When crop production (area and yield) is pushed to a maximum, to feed an ever increasing population, then the increased production from a 'good' year will not balance the deficit from a 'bad' year.

Chapter 18

CLIMATE AND INDUSTRY, COMMUNICATIONS AND TRANSPORT

18.1 Introduction

Climate plays such a large part in the process of everyday living that it is extremely difficult to cite just a few examples without doing an injustice to the other aspects. However, it has been decided that after the consideration given to the other factors in the previous chapters the main areas not so far considered are commerce, industry, transport, power and communication. The intricacies of the climatic applications in these sections are legion and we can do no more here than outline some of the ways in which climate plays a part in these activities.

18.2 Industry and Commerce

When the need for a new industry or commercial venture is realized and accepted there are numerous decisions to be made before the plan comes to fruition and many of these decisions depend for their solution on certain climatic variables.[1] For instance, the siting of a new industrial plant involves the consideration of the following points.

(1) Labour availability. The proximity of sources of labour depends, to a large degree, upon the nearness of large towns which have often been positioned for reasons in which climate played a part.

(2) Raw materials. Generally, the on-the-spot availability of some raw material is not a function of the climate except for such items as lumber camps, wood-pulping factories and such. It is certainly very necessary to have sufficient com-

munications with the sources, and this can depend upon the waterways, railways and roads (18.4).

(3) Cost of land. Normally, if the land is not too useful for agricultural purposes it can be had at a more reasonable cost in a newly developing area.

(4) Buildings. The erection of these and the type needed for the job on hand and the climatic region have been mentioned in Chapter 14.

(5) Transportation. Can the factory guarantee that the labour force will be able to commute easily; is public and private transport sufficient for its needs?

Others aspects, dependent upon meteorological factors, are the water supply: is this sufficient for the needs of the new industry; air pollution: will there be any problems concerning the disposal of waste products to the air; heating and cooling: will there be a great need at any time of the year to supply an exorbitant amount of heating and/or cooling for the comfort of the employees or the safe storage and keeping of the products? The storage, warehousing and transportation of the raw, intermediate or end products naturally raises problems dependent upon the climatic conditions of the region.

Workers' comfort will vary with climatic conditions and even when the buildings have been sensibly and efficiently constructed there is often need for artificial cooling and heating, thus adding greatly to the overhead expenses. Outdoor industries are, of course, even more at the mercy of the climate so that after the determination of the extremes of climatic conditions under which operations cannot proceed[2] (Fig. 18.1) it is essen-

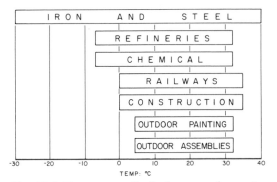

Fig. 18.1 Temperature limits for normal operating conditions

tial to find out for what percentage of the time these may exist at the site chosen.[3,4] For some operations there exists a linear relationship between cost or output and the temperature below certain thresholds.

Optimal conditions have also been worked out for some indoor occupations,[5] the climatic conditions dominant in these being temperature, humidity, light intensity and air flow. The use of the natural climate to its maximum extent can save a great deal of money by avoiding the high cost of man-made climate.

When the product is ready and available it is necessary to study the finer points of marketing, such as advertising. This also needs careful advance planning because a full-scale advertising campaign cannot be presented to the public overnight. Thus, if the producer wishes to know when is the optimal time for advertising anti-freeze in an area, the statistic can be estimated with a certain degree of probability from prior climatological knowledge.[6] Also, the sale of, say, swimming accessories, woollen clothing, sandals, ice cream and so on, will have a better likelihood of success if the time is optimal.

Mention has been made earlier (16.4) of the problem of atmospheric pollution and a whole book could easily be devoted to this aspect.[7,8] However, a few general comments may be made. Chemical effluents, when emitted into an atmosphere of unstable air and strong winds, will soon become so diminished in concentration that there will be little danger of seriously affecting vegetation, animals or humans. Under conditions of strong inversions, such as are often present on foggy days when the air is still, the effluents be-

come more and more concentrated since they are held in a smaller volume. These contaminants act as nuclei tending to give rise to small water-droplets causing the fog to become dense and laden with chemical components, such as sulphur dioxide. In a short time this can lead to serious respiratory complaints and people with cardio-respiratory trouble have been known to die from the effects of contaminants in these smogs. Once the smog has become pronounced, it is not until there is a marked change in the weather conditions that any relief can be expected. Some attempts have been made to classify regions according to their 'air pollution potential', based on available climatological data.[9,10] The effect of sunshine on the top of the smog layer can also be harmful for it may help to form new compounds, some of which may be more serious than the originals. The idea of creating smokeless areas is the only certain cure for preventing such conditions and this has been done in certain areas in the United Kingdom, such as selected London boroughs.

Hotels and holiday resorts, of course, are also affected by adverse weather conditions. Regions within the British Isles that have a greater than average amount of sunshine make quite a capital out of this, and if one year they suffer a poor summer then business drops. The winter sports resorts rely upon good snow coverage for a certain period of the year and they are also subject to the variations of the weather. It is interesting to note that in Europe it is only in the region around the Mediterranean that one can be fairly certain of benefiting from a great amount of sunshine during the summer, while the rest of the continent has rain either fairly uniformly distributed throughout the year or a marked maximum in the summer.

A recent application, which shows how wide is the scope of applied meteorology, is discussed by Houseman.[11] The erection of off-shore oil rigs in the Persian Gulf requires 48 hours of calm weather. In 1956 a violent storm caused loss of life and nearly $1 million in financial loss. Nowadays meteorological consultants are employed to forecast weather conditions in the vicinity of rigs.

18.3 Power and Communications

Various elements of the weather can act as sources of power; of these the three most important are radiation, wind flow and precipitation. We are all familiar nowadays with the idea of solar batteries used in satellites but not many people appreciate that these could be supplied for use at the surface, although, being at the mercy of varying radiation they would not prove as efficient or reliable. The sun's heat can also be used for heating or cooling houses (14.6), heating water, and by the use of reflectors in the form of parabolic mirrors can cook by radiation such foods as steaks and sausages.

The sun has also given the heat energy that is present in coal, peat, oil and natural gas. The distribution of these to those areas where they are in greatest demand is determined by the climatic requirements, and sufficient warning must be given if the demand is to be supplied.

The wind, through its pressure effect, can also be used as a source of natural power, as windmills demonstrate. The resulting energy may be expressed as[12]

$$P \text{ (kilowatts)} = 2 \times 10^6 \, av^3,$$

where a = exposed surface area in square metres and v = wind speed in km/hour.

For effective and economic operation there must be a steady wind above a certain threshold which, in practice, is about 30 km/hour for over 40 per cent of the time. Under such conditions problems of turbulence and icing then arise in some sites.

Precipitation is the basic element in such power machines as watermills and hydro-electricity supplies. At least a part of the massive amount of energy associated with the rainfall from a single storm should, wherever possible, be trapped and put to work.

Both power and communications in many parts of the world are transmitted by overhead cables. Such cables are subject to stress due to glaze storms, lightning and wind, being especially serious if the high winds and glazing occur together.[13] During periods of high temperature the wires expand and appreciable sagging occurs. Temperature variations also influence transmission and the functioning of apparatus, such as switches, insulators and transformers. Moisture also can affect adversely certain pieces of electrical apparatus. Radio reception is also affected both by the action of the weather on the antennae and by fluctuations caused in the atmospheric conditions, such as static, and the changes in the refraction of the waves.

Underground cabling would reduce a number of the effects on power transmission, provided the cables were completely waterproof, since the climate underground is remarkably uniform. The cost would be extremely high and it is really a question of economics as to whether it is better to pay for installation or continuous maintenance. Davies[14] in a discussion on the effect of temperature, wind and illumination on electricity demand, shows how the weather parameters can give the control engineer of the Grid Centre important indications of the likely demand.

18.4 Transport

The transport system of a region is usually adapted to cope with the normal climatic conditions, and because of this it is generally the extremes of weather that cause concern. Therefore we are dealing here with the shorter period phenomenon of the weather variations and not so much with the climatic conditions.

The extremes of weather often cause much inconvenience and, unfortunately, bring about a great increase in the accident rate. The most pronounced effects of the weather on transport are those caused by ice formation, intense atmospheric turbulence, torrential rains, poor visibility and, additionally, through such indirect effects as the weathering of materials and the modification of lubricants.

Transport can best be considered under four main headings, namely air, water, rail and road.

18.4.1 Air

In many countries the meteorological service has been able to obtain the money needed for growth because it renders many services to aviation, both civil and military. The fairly obvious rewards of good forecasting in aviation problems have led to stress being laid upon the aviation forecasting aspect of meteorology. Many meteorological elements play major roles in aviation and it is convenient to consider their parts under the two headings of 'terminal' and 'on route'.

The terminal or airport problems begin with

the decision concerning the siting of the airport. This question is one that is related to climatology for it involves long period studies pertaining to such elements as occurrence of fogs, low ceilings and allied phenomena. Information concerning the incidence of wind speed and direction, cloud ceiling and visibility is fundamental to the successful operation of an airport and it seems strange that locations are sometimes chosen that are anything but ideal. Interrelationships among the elements are often more important, such as the incidence of low ceilings with winds from a certain direction.[15] All airports should be in possession of climatological information of this type analysed on both seasonal and diurnal scales, for the latter are often taken into account when planning flight schedules.

The direction or orientation of the runway is decided by the prevailing wind directions so that the aircraft may land and take off approximately into the wind. Runway air temperatures are also needed for use in the calculation of life and cargo-carrying capacity of aircraft.

On route the main meteorological elements may be considered as wind velocity, turbulence, clouds, icing and thunderstorms. Knowledge of all these is essential to complete aircraft safety. Even with the latest civil and military aircraft equipped with radar which fly above many of the vagaries of the weather, information on these elements is needed for safe and economical operation.

The routing of a flight should take full cognizance of tail winds and head-wind and optimum levels in order to obtain the most economical route which also avoids areas of turbulence in which appreciable vertical movement and icing may occur.

18.4.2 Water

For thousands of years man has been aware of the effect of the weather upon transport by water of commodities and people. Even nowadays it is not uncommon to read of heavily loaded ferry boats being lost in high seas and the number of lives lost at sea due to inclement weather must surely be comparable with the number sacrificed in war.

At the time of the sailing ship man was especially at the mercy of the winds and a very ancient

application of climatological knowledge has been preserved showing how such information can be turned to economic advantage.

As was described earlier (Chapter 1), Hippalus used knowledge of the monsoonal reversals of wind direction in the Arabian sea to schedule sailing times, and also to save paying customs duty. Even the voyages of Ulysses have been interpreted in the light of information on the seasonal winds of the Mediterranean Sea. Nowadays it matters less to large vessels whether there is a tail or a head wind, although the latter has a greater effect than the former. Small craft must take care when strong winds occur suddenly; coastal and lake craft which are not generally equipped for severe weather need to pay especial attention to forecasts of gale force winds. The high speed winds associated with hurricanes and typhoons can cause damage to or complete loss of even large ships and warning is essential when such phenomena are expected in the vicinity.

Ice presents an appreciable hazard to shipping, particularly the so-called 'black' ice that deposits on all wind-chilled exposed surfaces, often occurring at such a rate that a small vessel is hard pressed to maintain an even keel. Many of the world's waterways are frozen for some period of the year and it is necessary to utilize climatological knowledge to forecast the likely freezing and thawing dates of these waterways. Information of this type has an obvious economic importance. In and around the Great Lakes of North America it has been discovered that the 'break-up' date is closely correlated with the mean temperature for February. In areas or seasons in which the ice-bound period is short, ice-breakers may be used to maintain a permanently open channel.

Harbours are normally chosen so as to provide protection from weather hazards, especially the high winds and high seas that may strike the anchored ships before they have a chance to ride out the storm. Fog is a special danger in harbours, particularly those situated in or near large cities where smog may occur.

Weather conditions are extremely important during the loading and unloading of cargoes for some commodities can be damaged or ruined by exposure to the climate elements. The transportation of perishables is also affected by the weather[16] for this factor decides whether heating or refrigeration must be used. A recent approach to the problem of storage aboard ships is to use cold sea water to chill the storage holds and so avoid the need for expensive refrigeration apparatus.

On inland waterways, such as canals, a drought spell can often cause such a reduction in water flow that traffic is reduced or even brought to a complete halt. This occurred in Europe during the summer of 1964.

18.4.3 *Railways*

The railway was instrumental in opening up many areas and countries to further and rapid development before the advent of the aeroplane. The examples of East Africa, Ecuador and parts of North-West India come to mind. The engineers concerned had to battle with many problems, climatic ones included, and the laying of a track appears to instil in people a confidence still not existing with aeroplane, road or sea travel. When severe conditions of icing and winds occur many travellers desert the other media and travel by railway, taking confidence in its 'permanent way'. This extra load, of both passengers and freight, adds to the problems of safe operation under extreme conditions. During times of climatic stress the tracks, signals and bridges are all subject to climatic elements, and additional complications may arise when the communications network itself is affected.[17] The railway, which plays such an important part of the transport system of the United Kingdom, is in most cases completely exposed to the elements. Floods, deep snow, earthslides, low visibility and very low temperatures can all play a part in creating havoc on the railway system. In the United Kingdom the first three parameters are fortunately very rare occurrences although in the extreme north of Scotland snow ploughs are kept busy during the winter months. Low visibility and temperatures are much more common but they still continue to cause a disproportionate amount of chaos, especially in and around the large cities where the railways serve a substantial commuting clientele. The five factors mentioned above give rise to some effects that can be reduced or avoided by sensible practices, such as snow fences, supported cuttings, heated points, etc., but really extreme conditions cannot be guarded against completely. In many mountainous areas such as the Alps and the Canadian Rockies, protec-

tion from avalanches of snow and/or earth is a very important item necessitating sensible use and interpretation of available climatic data.[18]

Good weather can also attract more passengers to the railways and lead to overcrowding of some of the antiquated systems, but its greatest effect is upon commodity transportation. The United Kingdom with its well-stocked waters, agricultural lands and large clusters of population has great need of swift means of transportation. Farm and sea crops must be brought to centres of consumer demand within the shortest reasonable time-limits, given distance, perishability of the commodity and so on. During warm weather it is essential either to provide refrigeration or to decrease the time of delivery, and railway management must be aware of the varying demands of the freight market and be able to cope with these.

18.4.4 *Roads*

Weather affects roadways throughout their existence, construction, maintenance and operation all being subject to atmospheric vagaries. At the time of construction the correct foundationing and material should be used so as to counteract the effects of the weather in the best possible way. For example, in very cold regions the severe freezing makes it imprudent to use a concrete road and many parts of Russia have roads made of sawn-off logs inserted vertically, so as to reduce the heaving effect that occurs during the seasonal variations. Again, the surfaces of some roads cannot stand the extreme summertime variations of surface temperatures which cause pronounced expansion and contraction that leads to cracks and wearing. Maintenance cost is then increased on such roads, as it is on the unsurfaced roads if these are to be kept open to traffic during the rainy seasons.

During bad weather it is the safety factors that are most reduced; poor visibility lowers safe travelling speed, slippery surfaces call for especially careful driving and extreme high winds may even blow cars from the roads in exposed places.[19]

Good weather generally leads to a great increase in the number of road users so that traffic congestion, loss of tempers and more accidents result. Many tarmacadam roads display pronounced melting of the surface when air temperatures rise above about 27 °C. As mentioned for the railways hot weather can give rise to problems for the haulers of fresh produce.

So far we have not mentioned the effect of weather upon the pedestrian. During heavy precipitation the gutters often become unable to cope with the flow of water so that crossing the road calls for agility in order to avoid the large pools. When snow falls in a built-up area the pavements often become dangerously slippery unless the snow is soon removed. When the thaw occurs, the roads and pavements become a sea of slush because of bad construction, planning and maintenance. During the summer-time the glare from the dazzling white pavements can be most pronounced, the rays reaching the eyes from a direction that is not protected by human physiology or the normal-sized sunglasses. The use of coloured pavements, such as green or pink, would reduce this annoyance whilst leading to an increased absorption of sunshine, especially useful in winter for melting snow.

18.5 Military Operations

The weather has influenced military action, as it has all the other endeavours which man has undertaken. The tactics of individual battles, and the strategy of wars have often been determined by the climate of the area. Some military planners have attempted to consider the effect of the climatic factors, but others have chosen to ignore this problem, either because they did not appreciate the weather factor, or because suitable information was not available to them.

Alexander the Great realized the need for information on the weather and included several advisers on his staff. Unfortunately, for both Alexander and the science of military climatology, the advisers often could not agree. Alexander was successful in his campaigns, either with, or in spite of, their advice.

Napoleon had many troubles with the inclement winter during his Russian campaign. Perhaps a better knowledge of the meteorological conditions with which his troops had to contend, would have caused him to alter his plans. Again, at the Battle of Waterloo, the heavy rains and flooding streams reduced the mobility of his troops. Many military analysts believe that on a dry field he would have carried the day.

The loss of the Spanish Armada marked the downfall of Spain as a sea power. While it is true that the English fleet destroyed some of the ships the greatest loss was caused by storms as the Armada attempted to round Scotland on the way back to Spain. Many other similar events could be retold.

In more recent history, military operations have made use of the elements. For the attack on Pearl Harbour the Japanese fleet advanced with a storm which was moving across the Pacific. The cloud cover was sufficient to prevent detection until they made their strike. The Japanese also made use of the weather in several other ways. One was to release balloons in Japan which were to be carried by the winds of the prevailing westerlies over the USA. The balloons carried incendiary bombs designed to start fires. Although the damage to the American war effort was slight, in relation to the risk and effort expended, the weapon was successful.

During the Second World War the Allies used weather factors in planning for the invasions of Europe. In the case of the Normandy invasion, the German meteorologists had indicated that the weather would not be suitable for invasion during the period, so the German troops were caught with their defences relaxed.

The requirements for the Normandy invasion were that D-Day should fall in the period one day before to two days after a new or full moon, that for three days after D-Day wind speeds should be less than 20 km/hour on shore and less than 40 km/hour at sea, that cloud base should be above 1 000 m with less than 3/10 cloud below 2 500 m and that visibility should be over 5 km. Climatological knowledge of the area showed that chances of such conditions were better in June (1 in 13) than in May (1 in 24) or July (1 in 33).[20]

Before starting the discussions of applied climatology to modern military usage there are three points which must be stressed. First, pure climatology does not give the type of information required. In fact classic climatological data are always insufficient and sometimes have been extremely misleading. Secondly, the planning information supplied by applied climatology should not be used as a short-range operational forecast. Thirdly, close co-operation between the military planner and the military climatologist is always necessary to ensure complete understanding.

Modern warfare is a complex operation with many facets. Suitable weather conditions for a certain tactic would be undesirable for another. For example, a heavy fog could cancel an air strafing attack but could provide cover to a commando-type raid. Some types of military operations are listed below and the weather information which the planner must know so as to provide for strength of reserve or diversified support are given. The list is not intended to be all-inclusive, nor to indicate the relative importance of an operation to any of the others.

Aircraft operations. For the safe operation of any aircraft, minimum visibility and ceiling conditions do exist, which together with strong winds, and certain hydrometeors, restrict use of an airfield. The military planner must know how frequently the base can be used. This may limit the type of aircraft and navigation aid available.

Tactical ground support. Here the planner relies on applied climatology to indicate the fraction of the day that air support can be expected. Now the problem increases because both the airfield and the support area must be considered together, and both must have suitable conditions concurrently. The weather at the airbases of the enemy must also be considered.

Strategic bombing. The weather requirements are usually more flexible in this operation. If visual bombing is not required, then the problem is further reduced. Again, the aircraft must take off and land but the distance is usually such that the aircraft may depart from widely separated airfields and suitable bases may be found for recovery. The chances of a successful mission are naturally reduced if the bombing is to be done visually. Planning a visual bombing raid calls for a study of secondary targets which may be attacked if the primary one is not visible.

Air reconnaissance. This is usually accomplished by means of photography, often from high altitudes. The planner must know how much of the time he will be deprived of reconnaissance aid. As in strategic bombing, other suitable air-bases can be used. If not, then the conditions at the base must also be included.

Amphibious assault. In addition to air-cover support, as listed above, the conditions of the sea become important. The surf, tides, tempera-ture, visibility and winds enter into the problem of landing on the beaches, and of continual supply after the landing.

Air-borne assault. For paratroopers, helicopters and gliders the wind conditions become critical. Often other conditions are also imposed, such as cloud cover, limiting the time available for such operations.

Soil conditions. Soil trafficability must be considered for overland passage. The composition of the soil is fixed, but the water content is extremely variable, and depends upon the nature of the rainfall, evaporation, drainage and soil itself.

Chemical warfare. Smoke and gases are carried by the wind and under some conditions disperse rapidly. The moisture content of the air also affects many agents. For incendiaries to be most effective the surface must be dry, indicating several days without rain. The climatologist must supply data on winds, stability, moisture, temperature and hydrometeors. Microclimate becomes more important in chemical warfare than in any other military operation.

Clothing and equipment. The adaptation of clothing to climate has been discussed earlier in the text. However, while the civilian can usually find some shelter from the elements, the soldier is often denied this shelter by the enemy. Trenchfoot and other similar ailments beset the soldier. Much attention must be given to the effect of combined elements, such as chill factors, heat exhaustion, the problem of keeping dry and warm, all under severe conditions. Also included is consideration of survival in extreme conditions, and equipment must be given the same testing. In these studies mean values lose their importance and the extremes must be considered. If, for example, a plastic item breaks on one cold night it is useless thereafter. The value of climatic extremes has been appreciated by the US Defense Department which has published a list of the extremes of various environmental factors to be anticipated in different climatic zones.[21]

Hazards. As in the case of the Spanish Armada, a major storm can wipe out armies, fleets and military installations. A hurricane, or ice storm leaves the military unit weakened and subject to attack. The military planner must be informed of all hazards which do occur in his theatre of operation so that proper allowances may be made for their infrequent but devastating occurrence.

The future makes the challenge even more difficult. Satellites, missiles and high-performance aircraft increase the importance of the weather, and make the problem more complex.

18.6 Miscellaneous

This, the final descriptive section of the book, attempts to collect together many of those aspects of life in which climate plays a role but for which no suitable section heading has arisen. In some ways it is a haphazard collection but it is necessary to present such an agglomeration rather than omit it completely.

In a climatic zone such as that in which the British Isles are situated the weather is a real hazard to all forms of outdoor entertainment. Especially feared is the ubiquitous rain: to such activities as open-air theatres and garden fêtes even a short duration fall can bring great disappointment while, if it is prolonged, sports such as bowls, cricket and tennis have to be abandoned. It is possible in some cases to choose, by using a climatic summary, a period when the rainfall probability is least — such appears to have occurred fortuitously in the case of the selection of the time of the Wimbledon Lawn Tennis Championships, although during the past few years the luck has not held so well. In many cases insurance is now taken out in order to compensate for an abandonment due to inclement weather.

Many materials exposed to the elements show some form of weathering; for example, paints and coloured materials fade, iron-containing metals can rust, wood can warp or rot. Even plastics can degenerate when subjected to normal climatic conditions.[22] Winds carrying much dust or sand can penetrate exposed equipment and thereby lead to the clogging of machines, vents, etc., as well as abrasing external surfaces. The subject of weathering of materials is an extremely wide one and it is advisable that the person interested in the behaviour of certain materials under given conditions carry out his own experiments. Nowadays some commodities stress their special modifications for a certain area, such as paints adapted to the local climatic conditions.

Hobbies, such as numismatism, philately and photography, to name but a few, are also affected by the weather. Coins should be protected from damp conditions, an environment that can also play havoc with mint stamps if the gum begins to soften. In photography the appearance of a subject, animate or inanimate, will change with the lighting conditions. Many disappointments have occurred when the photographer has set his camera for one set of light values only to have them change radically by the time he actually takes the shot. The behaviour of the film itself changes with conditions, especially temperature.

In the light of our climatic knowledge, we may now think that men of old were not so far wrong, at least in order of importance and emphasis, when they called the weather factors, such as thunder, lightning, rain and winds, gods, for indeed it is these factors which govern the outward pattern of our lives, as it hoped has been shown in this book.

REFERENCES

Chapter 1

1. 'The Business of Weather', *Bull. Amer. Met. Soc.* **44**, 63–79, 1963, 4 papers.
2. M. I. Budyko, *Climate and Life,* Academic Press, 1974.
3. D. M. Gates, *Man and his Environment: Climate,* Harper and Row, 1972.
4. W. P. Lowry, *Weather and Life,* Academic Press, 1969.
5. J. R. Mather, *Climatology, Fundamentals and Applications,* McGraw-Hill, 1974.
6. W. J. Maunder, *The Value of the Weather,* Methuen, 1970.
7. J. E. Oliver, *Climate and Man's Environment,* Wiley, 1973.

Chapter 2

1. R. B. Platt and J. F. Griffiths, *Environmental Measurement and Interpretation,* Krieger, 1972.
2. J. L. Monteith, *Survey of Instruments for Micrometeorology,* Blackwell, 1972.
3. W. E. K. Middleton and A. F. Spilhaus, *Meteorological Instruments,* Univ. of Toronto Press, 1965.
4. HMSO, *Handbook of Meteorological Instruments,* Vol. **1** (1956), Vol. 2 (1961), London, England.

Chapter 3

1. J. Glover and J. S. G. McCulloch, *Q.J.Roy. Met. Soc.* **84**, 172, 1958.
2. D. Brunt, *Physical and Dynamical Meteorology,* Camb. Univ. Pr., 1941.
3. G. L. Barger, R. H. Shaw and R. F. Dale, *Agric. and Home Econ. Expt. Sta. Bull.,* Iowa State Univ., Dec. 1959.
4. B. W. Thompson, *Tech. Mem., E. Afr. Met. Dept.* **8**, 1957.
5. R. E. Huschke (ed.), *Glossary of Meteorology,* Amer. Met. Soc., 1959.

6. W. Palmer, *5th Nat. Conf. Agric. Met.,* Lakeland, Florida, Apr. 1963.
7. R. Geiger, *The Climate near the Ground,* Harvard Univ. Press, 1965.
8. E. L. Deacon, *Q.J.Roy. Met. Soc.* **75**, 89, 1949.
9. H. L. Penman, *Q.J.Roy. Met. Soc.* **75**, 293, 1949.

Chapter 4

1. D. Brunt, *Physical and Dynamical Meteorology,* Camb. Univ. Pr., 1941.
2. H. R. Byers, *General Meteorology,* McGraw-Hill, 1974.
3. R. E. Huschke (ed.), *Glossary of Meteorology,* Amer. Met. Soc., 1959.

Chapter 6

1. J. Gentilli, *Geography of Climate,* Univ. of W. Australia Press, 1958.
2. W. Köppen, *Grundriss der Klimakunde,* Berlin and Leipzig, 1931.
3. C. W. Thornthwaite, *Geog. Rev.* **38**, 55, 1948.
4. L. R. Holdridge, *Life Zone Ecology,* Tropical Science Centre, 1957.

Chapter 7

1. P. E. Waggoner, *Bull.* **656**, Connecticut Agr. Expt. Stn, Apr. 1963.
2. R. Geiger, *The Climate Near the Ground,* Harvard Univ. Pr., 1965.
3. E. Brezina and W. Schmidt, *Das künstliche Klima in der Umgebung des Menschen,* Stuttgart 1937.
4. N. T. Rosenberg, *5th Nat. Conf. Agric. Met.,* Lakeland, Florida, Apr. 1963.
5. M. Jensen, *Shelter Effect,* Danish Tech. Pr., 1954.
6. J. H. Chang, *Climate and Agriculture,* Aldine, 1968.
7. B. J. Mason, *The Physics of Clouds,* Clarendon Press, 1957.

8. V. K. La Mer, *Retardation of Evaporation by Monolayers,* Academic Press, 1962.

Chapter 8

1. N. K. Horrocks, *Physical Geography and Climatology,* Longmans, 1961.
2. W. H. Wischmeier, *Proc. Soil Sci. Soc. Amer.* 23, 246, 1959.
3. W. H. Wischmeier and D. D. Smith, *Trans. Amer. Geophys. Un.* 39, 285, 1958.
4. *FAO Agricultural Development Paper,* 71, 1960.

Chapter 9

1. A. Vernet, *UNESCO Arid Zone Research* X, 75, 1958.
2. M. I. Newbigin, *Plant and Animal Geography,* Methuen, 1960.

Chapter 10

1. W. M. Parker, *Soil Sci.* 62, 109, 1946.
2. P. M. A. Bourke, *Biometeorology,* Pergamon, 1962, p. 153.
3. C. P. Willsie, *Crop Adaptation and Distribution,* San Francisco, 1962.
4. J. H. Chang, *Climate and Agriculture,* Aldine, 1960.
5. M. H. Kimball and F. A. Brooks, *Calif. Agric.* 13(5), 7, 1959.
6. N. A. Maximov, *Dratkii Kurs Fiziologii Rastenii* (*A Short Course on Plant Physiology*), Moscow, 1948.
7. A. D. Hopkins, *US Dept. Agric., Misc. Publ.* 280, 1938.
8. M. Y. Nuttonson, in *Vernalization and Photoperiodism,* ed. A. E. Murneek, Ronald Press, 1948.
9. D. M. Brown, *5th Nat. Conf. Agric. Met.,* Lakeland, Florida, Apr. 1963.
10. W. P. Lowry, *Weather and Life,* Academic Press, 1970.
11. H. Geslin, *Congrès Intern. Pedologie,* Montpelier, France, Paper 32, 1947.
12. E. Guyot, *Publ. Inst. Met., Univ. Thessaloniki* 5, 1956.
13. J. R. Hildreth and E. Burnett, *Proc. Conf. on Weather: Agric. Res. in the Great Plains,* USWB and USDA, 1953, p. 23.
14. H. H. Laude, *ibid.,* p. 41.
15. D. Berenyi, *Acta Universitatis Debrecensis* 3, 229, 1957.
16. G. D. V. Williams and G. W. Robertson, *Can. J. Pl. Sci.,* 45, 34, 1964.
17. J. C. Das, Meteor. Dept., Lusaka, June 1973.
18. W. Baier, *Int. J. Bioclim. Biomet.* 17, 313, 1973.
19. J. Y. Wang, *Agricultural Meteorology,* Pacemaker Press, 1963.

20. A. T. Selianinov, *Vsesoiuznoe Geograficheskoe Obshchestvo, Leningrad, Izvestia,* 89, 225, 1957.
21. R. M. Whittle and W. J. C. Lawrence, *J. Agric. Eng. Res.* 4, 326, 1959; 5, 36, 165 & 399, 1960.
22. A. A. Jackson, *Weather, London,* 14, 117, 155, 1959.

Chapter 11

1. S. S. Paterson, *Goteborg, Universitet, Geografiska Institut, Meddelande,* 51, 1956.
2. F. Lauscher, *Wetter and Leben,* 12, 98, 1960.
3. L. R. Holdridge, *Life Zone Ecology,* Tropical Science Center, Costa Rica, 1967.
4. R. Geiger, *The Climate Near the Ground,* Harvard University Press, 1965.
5. H. Mrose, *Angew, Met.* 281, 1956.
6. I. Fel'dman, *Priroda, Moscow* 5, 93, 1959.
7. V. N. Kaulin, Effect of forest belts in Kamennaia Steppe on precipitation, *Meteorologiia i Gidrologiia,* 6, 32, 1962.
8. J. Schubert, *Zeit. F. Forst. U. Jagdw.* 69, 605, 1937.
9. E. Trapp, *Biokl. B,* 5, 153, 1938.
10. J. Deinhofer and E. Lauscher, *Met. Zeit.* 56, 153, 1939.
11. K. Krenn, *Wien Allg. For. U. Jagdztg.,* 1933.
12. J. F. Griffiths, *Proc. CCTA Conf. Housing and Urbanization,* Nairobi, 1959.
13. R. Geiger, *Forstw. Central, Berlin* 48, 337, 1926.
14. R. Geiger and H. Amann, *Forstw. Central, Berlin* 53, 237, 1931.
15. J. F. Nagel, *Q.J.Roy. Met. Soc.* 82, 452, 1956.
16. A. L. Griffith, *VIIth Brit. Commonw. For. Conf., Austr. and N.Z.* 2, 1957.
17. H. C. Fritts, *Ecology,* 40, 261, 1959.
18. F. P. Keen, *M. Weath. Rev.* 65, 175, 1937.
19. S. J. Curry and J. F. Griffiths, *Ecology* 40, 490, 1959.
20. R. A. Read, *Windbreaks for the Central Great Plains,* Nebraska Agric. Expt., 1965.
21. World Meteorological Organisation, *Windbreaks and Shelterbelts,* Publ. 147, Geneva, 1964.

Chapter 12

1. E. Huntington, *Civilization and Climate,* Yale Univ. Pr., 1924.
2. S. F. Markham, *Climate and the Energy of Nations,* Oxford Univ. Press, 1947.
3. A. P. Gagge, G. E. A. Winslow, L. P. Herrington, *Amer. J. Physiol.* 124, 30, 1938.
4. G. Manley, *Geographical Rev.* 48, 98, 1958.
5. J. B. Rigg, *Weather* 16, 255, 298, 327, 1961.
6. E. F. Adolph, *Physiology of Man in the Desert,* Interscience, 1947.

7. T. Bedford, *Medical Res. Counc. War Memo,* **17**, 1946.
8. P. A. Siple and C. F. Passel, *Proc. Amer. Phil. Soc.* **89**, 177, 1945.
9. A Court, *Bull. Amer. Met. Soc.* **29**, 487, 1948.
10. R. G. Steadman, *J. Appl. Met.* **10**, 674, 1971.
11. G. Galleotti, *Pflug. Arch. ges. Physiol.* **160**, 27, 1914.
12. F. C. Houghten and C. P. Yaglou, *Trans. Amer. Soc. Heating and Vent., Eng.* **29**, 193, 1923.
13. E. C. Thom, *Air Cond., Heating and Vent.* June 1957.
14. D. H. K. Lee and A. Henschel, *Ann. N.Y. Acad. Sci.,* **134**, 1966.
15. H. E. Landsberg, *World Meteor. Orgn. Publ. 331,* Geneva, 1972.
16. P. A. Siple, *Physiology of Heat Regulation and Science of Clothing,* L. M. Newburgh (ed.), Saunders (Philadelphia), 1949, Ch. 9.
17. E. K. Hearn, *Home Economics,* Nov. 1961.
18. T. B. Franklin, *Climates in Miniature,* Philosophical Library, 1955.
19. S. W. Tromp, *Medical Biometeorology,* Elsevier, 1963.
20. S. Licht, *Medical Climatology,* E. Licht., New Haven, Conn., 1964.
21. C. A. Mills, *Climate Makes the Man,* Harper Bros., 1942.
22. J. LeBlanc, *J. Appl. Physiol.* **17**, 950, 1962.
23. E. Schikele, *Military Surgeon* **100**, 235, 1947.
24. H. Brezewsky, *Morbidity and Weather,* in Licht (ed.), *Medical Climatology,* New Haven, Conn., Ch. 13, 1964.

Chapter 13
1. C. P. Luck and P. G. Wright, *J. Physiol.* **147**, 53, 1959.
2. K. Schmidt-Nielsen, B. Schmidt-Nielsen, S. A. Jarnum, T. R. Houpt, *Amer. J. Physiol,* **188**, 103, 1957.
3. K. Johansen, *Inter. J. Biomet.* **VI**, 3, 1962.
4. J. C. Bonsma, J. V. Marle and J. H. Hofmeyer, *Empire J. of Exp. Agric.* **XXI**, 83, 1953.
5. D. H. K. Lee, *FAO Development Paper* **38**, Dec. 1953.
6. J. D. Findlay, *Bull. Hannah Dairy Research Inst., Ayr* **9**, 1950.
7. S. Brody, *J. Dairy Sci.* **39**, 715, 1956.
8. P. D. Sturkie, *Avian Physiology,* Comstock (Ithaca, N.Y.), 1954.
9. Beacon Foods, *Profitable Poultry Management,* 24th ed., Cayuga, 1962.
10. D. A. Parry, *J. Exp. Biol.* **28**, 445, 1951.
11. R. Geiger, *The Climate Near the Ground,* Harvard Univ. Pr., 1965.

12. R. C. Rainey, *Q.J.Roy. Met. Soc.* **84**, 334, 1958.
13. M. Kleem, *Naturf.* **6**, 254, 1929.
14. P. A. Glick, *US Dept. Agric. Tech. Bull.* **673**, 1939.
15. W. O. Pruitt, *Arctic, Montreal* **10**, 131, 1957.
16. P. F. Scholander, R. J. Hock, V. Walters, F. Johnson and L. Irving, *Biol. Bull.* **98–99**, 237, 1950.
17. D. M. Gates, *Man and his Environment: Climate,* Harper & Row, 1973.

Chapter 14
1. L. J. Sutton, *Egyptian Ministry of Public Works, Phys. Dept. Paper* **50**, 1945.
2. P. A. Buxton, *J. Anim. Ecol.* **1**, 152, 1932.
3. S. Polli, *Trieste Publ.* **300**, *Alp Giulie* **52**, 22, 1953.
4. J. K. Page, *Internat. J. Bioclim. & Biomet.* **2** (4), 29, 1958.
5. J. K. Page, *Internat. J. Bioclim. & Biomet.* **8** (2), 97, 1964.
6. UNESCO., *Climate and House Design,* New York, 1971.
7. J. E. Aronin, *Climate and Architecture,* Reinhold, 1956.
8. E. N. van Deventer, *Nat. Bldg. Res. Inst., Pretoria, Bull.* **17**, 69, 1959.
9. P. A. Siple, *Bull. Amer. Inst. Arch.* Sept. 1956.
10. D. A. Thomas and J. B. Dick, *J. Inst. Heating and Vent. Eng.* **21**, 85, 1953.
11. B. H. Evans, *Texas Eng. Exp. Sta., Res. Rep.* **59**, March 1957.
12. H. C. S. Thom, *Proc. Amer. Soc. Civil Eng.* **80**, Nov. 1954.
13. P. Moon and D. E. Spencer, *Illum. Eng.* **37**, 707, 1942.
14. P. Petherbridge, *Geophys. Bull.* **10**, Oct. 1954.
15. R. E. Lacy, *Q.J.Roy. Met. Soc.* **77**, 283, 1951.
16. W. Thein, *Ann. Hydrogr.* **61**, 196, 1938.
17. W. E. K. Middleton and F. G. Millar, *J. Roy. Astron. Soc., Canada,* **30**, 265, 1936.
18. J. F. Griffiths, *Proc. CCTA Conf. Housing and Urbanization,* Nairobi, 1959.
19. H. E. Landsberg, *Physical Climatology,* Gray Publ. Co., 1960, p. 391.
20. W. Schmidt, *Meteorologische Feldversuche Uber Frostabwehrmittel,* Vienna, 1929.
21. B. Givoni, *Man, Climate and Architecture,* Elsevier, 1969.
22. J. Stein, *Smithsonian,* **4**, 29.
23. H. C. S. Thom, *Amer. Soc. of Heating, Ref. and Air Cond. Eng., Conf., Dallas,* 1960.
24. H. C. S. Thom, *ASHAE J., Sect., Heating, Piping, Air Cond.,* p. 137, 1956.
25. K. J. K. Buettner, *Biometeorology,* p. 128, Pergamon Press, 1963.

26. D. W. Boyd, *Proc. 29th Western Snow Conf., Spokane, Washington,* **6**, 1961.

Chapter 15

1. *US Weather Report,* 28th Feb. 1959.
2. R. K. Linsley, Jr., M. A. Kohler, J. L. H. Paulhus, *Applied Hydrology,* McGraw-Hill, 1949.
3. J. V. Sutcliffe, W. R. Rangley, *IASH Comm. of Surface Waters, Publ.* **51**, 182, 1960.
4. Ven Te Chow, *Open-Channel Hydraulics,* Mc-Graw-Hill, 1959.
5. W. K. Henry, *Weatherwise* **18**, 80, 1965.
6. USWB, *Tech. Papers* 24 (1954), 28 (1956), **29** (1959), **40** (1961), **43** (1962).
7. E. J. Gumbel, *J. Inst. Water Eng.* **12**, No. 3, May 1958.
8. A. F. Jenkinson, *Q.J.Roy. Met. Soc.* **81**, 348, 1955.

Chapter 16

1. T. R. Detwyler and M. G. Marcus (ed.), *Urbanization and Environment,* Duxbury Press, 1972.
2. J. F. Griffiths and M. J. Griffiths, N.O.A.A. Technical Memorandum, EDS 21, 1974.
3. A. Bryant, *The Medieval Foundation of England,* Collier, 1968.
4. World Meteorological Organization *Urban Climates,* No. 254, 1970.
5. Swedish State Institute for Building Research, *Teaching the Teachers* (3 volumes), 1972.
6. R. C. Runnels, D. Randerson and J. F. Griffiths, *Int. J. Biometeor.,* **16**, 119, 1972.
7. W. P. Lowry, *Weather and Life,* Academic Press, 1969.

Chapter 17

1. E. Dorf, *Amer. Scient.* 48, 341, 1960.
2. T. Alexander, *Fortune,* **89**, 90, 1974.
3. J. E. Oliver, *Climate and Environment,* Wiley, 1973.
4. W. D. Sellers, *Physical Climatology,* Chicago Univ. Pr., 1965.
5. M. I. Budyko, *Climate and Life,* Academic Press, 1974.
6. R. A. Bryson, A Perspective on Climatic Change, *Science* **184**, 753, 1974.
7. P. J. Hart, ed., *The Earth's Crust and Upper Mantle,* Geophys. Monograph 13, American Geophysical Union, 1969.
8. E. W. Barrett, *Sol. Energy,* **13**, 323, 1971.
9. G. N. Plass, *Scient. Am.* **201**, 41, 1959.
10. J. F. Griffiths and R. E. Harriman, *Int. J. Environmental Stud.* 5, 271, 1974.
11. J. M. Mitchell, *Ann. N.Y. Acad. Sci.,* **95**, 235, 1961.

12. G. S. Callendar, *Q.J.Roy. Met. Soc.* 87, 1, 1961.
13. E. B. Kraus, *Q.J.Roy. Met. Soc.* **81**, 98, 1955.
14. G. B. Tucker, *Weather* **19**, 302, 1964.
15. H. H. Lamb, What can we find out about the Trend of our Climate? *Weather* 18, 194, 1963.
16. C. E. P. Brooks, Geological and historical Aspects of climatic Change. *Compendium of Meteorology,* p. 1004, Amer. Met. Soc., 1951.

Chapter 18

1. US Weather Bureau, *Weather a Factor in Plant Location,* Washington D.C., 1961.
2. H. E. Landsberg, *Physical Climatology,* Gray Printing Co., 1960, p. 391.
3. J. D. McQuigg and W. L. Decker, *J. Appl. Met.* **1** (2), 178, 1962.
4. H. C. S. Thom, *ASHAE J., Sect., Heating, Piping, Air Cond.* p. 137, 1956.
5. G. Grundke, *Petermanns Geogr. Mittl.* **255**, 1955.
6. F. Linden, *Business Record,* p. 144, March, 1959.
7. H. P. Kramer, *Met. Abstr. and Bibliogr.* **1**, 119, 1950.
8. J. J. Christensen, *Proc. Amer. Ass. Adv. Sci.* **17**, 78, 1942.
9. C. R. Hosler, Low-Level Inversion Frequence in the Contiguous United States, *M.W.R.* **89**, 319, 1961.
10. R. J. Moore *et al., Wind & Weather Summaries for Chemical Plant Design and Air Pollution Control,* Amer. Soc. for Testing Materials, *Special Tech. Pub.* **281**, 1960.
11. J. Houseman, *Weather* **16** (5), 139, 1961.
12. H. E. Landsberg, *Proc. Amer. Ass. Adv. Sci.* **17**, 402, 1942.
13. K. Pochop, *Energetika,* Prague **10** (5), 250, 1960.
14. M. Davies, *Weather* **15** (1), 18, 1960.
15. W. K. Henry, *Ladd Air Force Base,* US Air Force (typescript), 1954.
16. N. L. Canfield, Moisture damage to ships' cargoes, *Proc. Nat. Conf. on Applied Met.,* Hartford, Conn., 1957.
17. W. W. Hay, *Meteor. Monog.,* **2**, 1957.
18. P. Schaerer, *Avalanche Defenses for the Trans-Canada Highway at Rogers Pass,* Nat. Res. Council of Canada, Div. of Building Res., *Tech. Paper* **14**, 1962.
19. W. Benner, *Amer. Met. Soc., Bull.* **39** (8), 421, 1958.
20. A. Baroni, *Riv. Met. Aeronaut.,* Rome **22**, 53, 1962.

21. US Asst. Sec. of Defense, Supply & Logistics, *Climatic Extremes for Military Equipment,* Washington, D.C., US Govt. Printing Office, 1957.

22. R. B. Barrett, *Resistance of Plastics to Outdoor Exposure,* US Piscatinny Arsenal, Dover, N.J., *Tech. Rep.* **2102**, 1955.

INDEX

air conditioning, 92, 96
air flow
    foehn winds, 18
    gravity, 18
    land and sea breezes, 18
    measurement of, 7
    variation with altitude, 19, 94
air masses, 23, 25
air pollution potential, 121
allergies, 83
animals, climatic effect on distribu-
    tion of, 85, 91
architecture, requirements from
    climatic data, 93
atmospheric composition, 114, 115

bats, 86
bioclimatics, 64
bioclimatology, definition of, 2
birds, 88, 89
black body, 11
blight, 62
body temperature
    of birds, 88
    of mammals, 85, 88
    of reptiles, 90
building
    orientation of, 95
    sites and climate, 95
    temperature lag in, 96
    wind pressure on, 94

camel, 85
carbon dioxide, 115
cargoes and weather, 123, 124
catchments, 102
cattle, 86, 87
cave climate, 92
chernozems, 55
chinook winds, 18
cleaning, 81
climate
    and crops, 60, 65
    cave, 92
    continental, 13, 35

effect of changes in, 118
    indoors, 96
    marine, 13, 35
    past, 34, 35
climatic changes
    occurring now, 115, 118
    effect of, 118
climatic classifications, 30, 35–38, 77
    78
    elements and crops, 60
    factors, 21, 26, 112
    variability, 111
climatic modification by
    artificial stimulation of rain, 51,
      52
    frost prevention, 53
    reduction of evaporation, 52
    windbreaks, 51
climatography, 1
climagram, 86
clothing,
    insulation of, 79
    world zones of, 77, 78
cloud
    and sunshine, 12, 13
    estimation of, 5
    seeding, 51, 52
comfort, cost of, 98
contrails, 115
cooking and altitude, 80
cooling degree day, 97
coriolis effect, 21

daylight factors in buildings, 94
degree days, 65, 97
depth–area–duration relationships,
    102, 103
deserts, 59
dew, 63
dew point, 17
diet, 80
discomfort index, 76
domestic animals, 86, 87
drizzle, definition of, 102

drought
    definition of, 16, 17
    forms of, 62
dust, 115
dust devils, 19

ecliptic, 113
effective radiation surface, 68
engineering hydrology, 104
epiphytes, 57
equation of time, 10
equinoxes, precession of, 113
evaporation
    amount of, 64
    calculation of, 19, 20
    measurement of, 8
    reduction of, 52
evapotranspiration
    calculation of, 20
    measurement of, 20
extreme value analysis, 105

field capacity, 62
flood plains, 93
floods, 105
foehn wind, 18
fog, definition of, 102
food, climatic effects on, 79
forest
    climate of, 67–71
    fires, 70, 71
    influence on precipitation, 67, 68
    productivity, 70
forests, 57, 58
fowl, 88, 89
fronts, 25
frost
    advection, 53, 61
    heave, 54
    hollow, 19, 70
    prevention of, 53
    radiation, 53, 61

gardening, 81
general circulation, 21, 23
glasshouses, 66
grasslands, 58
greenhouse effect, 115
groundwater flow, 104
growing season, 62

hail, 63
health, 81, 82
heat
    balance of man, 70–75
    load on man, 73
    metabolic, 73
    prickly, 82
heating degree day, 97
heating threshold, 97
hobbies, 127
holidays, 121

house, orientation of, 96
housing, classification of, 98–100
humidity
    instrument accuracy, 7
    instruments, 7
    parameters, 7, 17
humus, 54
hurricane, 25, 83
hydrologic cycle, 101
hygrophytes, 57
hythergraph, 86

icing hazards, 122, 123
indoor climate, 96
infra-red thermometers, 12
insects, 70, 71
    and behaviour, 89
    as vectors of disease, 83
    their distribution with height, 89
insulation, 78, 79
irrigation, 63
isophanes, 64
I.T.C.Z., 115

Kirchhoff's law, 12
Köppen's classification, 35, 36

land and sea breezes, 18
lapse rate, 14
laterization, 55
leaching, 54
life zone classification, 36–38
light
    and radiation, 11
    definition of, 5
    instruments, 5, 11
locusts, 88, 89
long day plants, 63

mammals, body temperature of, 85,
    88
man
    average, 73
    heat balance of, 72
    water balance of, 75
marketing and weather, 121
materials and weather, 127
mean radiant temperature, 73
mesophytes, 57
mesoscale disturbances, 25
metabolic heat, 73
microclimate, 49
microclimatic variations
    of airflow, 50
    of evaporation, 50
    of humidity, 50
    of light, 49
    of precipitation, 50
    of radiation, 49
    of sunshine, 49
    of temperature, 49, 50
military operations, 125, 126

monotremes, 85
moulds, 79
mountains, climatic zones on, 34, 35
mulch, 62

Normandy invasion, 125

ocean currents, 21
oil wells, weather effect on drilling of,
    121
optical air mass, 10, 11

pachyderms, 85
pedestrians and weather, 125
phenology, 64, 65
photoperiodism, 63
photosynthesis, 63
pigs, 88
Planck's law, 11
plant location, 120
podzolization, 55
podzols, 55
pollution, atmospheric, 121
precipitation
    altitude variation of, 16
    annual, 15, 16
    artificial stimulation of, 51, 52
    causes of, 15
    diurnal variation of, 16
    early records of, 6
    forest influence on, 67
    horizontal, 70, 102
    impaction and soil erosion, 56
    incidence angle of, 95
    instrument accuracy, 6
    instruments, 6, 7
    instrument shields, 6
    intensity variation, 103
    monthly, 16
    record falls of, 15
    variation, 16
pressure, 8
prickly heat, 82

radiation, 10
    balance, 11
    depletion and scattering, 10, 11
    increment, 79
    instrument accuracy, 3
    instruments, 3
    long wave, 11, 12
    short wave, 4
    window, 12
radiative heat load on man, 73
rainfall
    penetration of, 94, 95
    estimation of by radar, 7, 102
reptiles, 90
reservoir
    flood control of, 104
    storage, 104

road surface, 125
runways, 123

seeds, dispersal of, 64, 71
sheep, 87
shelterbelts, 51
short day plants, 63
shrubland, 58
sierozems, 55
smog, definition of, 82, 108, 110
smudge pots, 53
snow, 17
soil
    erosion, 56, 64
    moisture, 62
soils, 54—56
solar
    constant, 10, 114
    declination, 10
sport, 127
standard continent, 26
    and soil, 58
    and vegetation, 59
    annual precipitation of, 28
    precipitation patterns on, 27, 28
    temperature patterns, 26, 27
stormflow, 103
strain index, 77
sub-surface flow, 104
sunscald, 61
sunshine
    and cloud, 12, 13
    and radiation, 12
    instrument accuracy, 4, 5
    instruments, 4
sun spots, 117, 118
surface run-off, 103, 104
sweating rates, 75

temperature, 13—15
    altitude effect on, 14
    annual cycle of, 13
    diurnal cycle of, 13
    effective, 61, 76
    effects on crops, 61
    for sleeping, 96
    instrument accuracy, 5
    instruments, 5
    lapse rate, 115
    mean radiant, 73
    records, 13
    soil, 6, 14, 15, 90
    super adiabatic lapse of, 14
    terms, 13
    water, 90
temperature—humidity index, 76
thermal wind decrement, 78
Thiessen network, 103
Thornthwaite's classification, 35, 36
timber production, 70
transport, 122—125
trees, temperatures within, 71

tropophytes, 57
typhoons, 25, 83

urban
    climate, 107, 108
    climatic changes, 107, 109
    dust dome, 108
    heat island, 108
    hydrology, 106, 107, 110
    physical characteristics, 106
    pollution, 109, 110
    vs. rural climates, 108

vegetation and climate, 57–59
virga, 15, 102

weather modification, 51–53

wilting point, 62
windbreaks, 51, 71
wind chill factor, 74, 75
wind
    erosion, 56, 74
    power, 122
    pressure on structures, 94
wind speed,
    in pine stand, 69
    variation with altitude, 94
    variation with leaf density, 69
woodland, types of, 57, 58

xerophytes, 57

yeasts, 79